The Idiot Brain

Dr Dean Burnett is a neuroscientist working as a tutor and lecturer based at Cardiff University's Institute of Psychological Medicine and Clinical Neurosciences. He dabbles in stand-up comedy and writes a popular science blog, 'Brain Flapping', for the *Guardian*.

The Idiot Brain

A Neuroscientist Explains What Your Head is Really Up To

DEAN BURNETT

First published in 2016
by Guardian Books, Kings Place, 90 York Way, London, N1 9GU
and Faber & Faber Limited
Bloomsbury House, 74–77 Great Russell Street
London WC1B 3DA

A CIP record for this book
is available from the British Library

ISBN 978–1–78335–081–0

2 4 6 8 10 9 7 5 3 1

Dedicated to every human with a brain.
It's not an easy thing to put up with, so well done.

Contents

Introduction

This book begins the same way as nearly all my social interactions; with a series of detailed and thorough apologies.

Firstly, if you end up reading this book and not liking it, I'm sorry. It's impossible to produce something that will be liked by everyone. If I could do that, I'd be the democratically elected leader of the world by now. Or Dolly Parton.

To me, the subjects covered in this book, focusing on the weird and peculiar processes in the brain and the illogical behaviours they produce, are endlessly fascinating. For example, did you know that your memory is egotistical? You might think it's an accurate record of things that have happened to you or stuff you've learned, but it isn't. Your memory often tweaks and adjusts the information it stores to make you look better, like a doting mother pointing out how wonderful her little Timmy was in the school play, even though little Timmy just stood there, picking his nose and dribbling.

Or how about the fact that stress can actually *increase* your performance at a task? It's a neurological process, not just 'something people say'. Deadlines are one of the most common ways of inducing stress that provoke an increase in performance. If the latter chapters of this book suddenly improve in quality, you now know why.

Secondly, while this is technically a science book, if you were expecting a sober discussion of the brain and its workings, then I apologise. You won't be getting that. I am not from a 'traditional' scientific background; I'm the first out of everyone in my family to so much as think about going to university, let alone go, stay there and end up with a doctorate. It was these strange academic inclinations, so at odds with my closest relatives, that first got me into neuroscience and psychology, as I wondered, 'Why am I like this?' I never really found a satisfying answer, but I did develop a strong interest in the brain and its workings, as well as in science in general.

Science is the work of humans. By and large, humans are messy, chaotic and illogical creatures (due largely to the workings of the human brain) and much of science reflects this. Someone decided long ago that science writing should always be lofty and serious, and this notion seems to have stuck. Most of my professional life has been dedicated to challenging it, and this book is the latest expression of that.

Thirdly, I'd like to say sorry to any readers who find themselves referencing this book and subsequently losing an argument with a neuroscientist. In the world of brain sciences, our understanding changes all the time. For every claim or statement made in this book, you'd probably be able to find some new study or investigation that argues against it. But, for the benefit of any newcomers to science reading, this is pretty much always the case with any area of modern science.

Fourthly, if you feel the brain is a mysterious and ineffable object, some borderline-mystical construct, the bridge between the human experience and the realm of the unknown, etc., then I'm sorry; you're really not going to like this book.

Don't get me wrong, there really is nothing as baffling as the human brain; it is incredibly interesting. But there's also this bizarre impression that the brain is 'special', exempt from criticism, privileged in some way, and our understanding of it is so limited that we've barely scratched the surface of what it's capable of. With all due respect, this is nonsense.

The brain is still an internal organ in the human body, and as such is a tangled mess of habits, traits, outdated processes and inefficient systems. In many ways, the brain is a victim of its own success; it's evolved over millions of years to reach this current level of complexity, but as a result it has accrued a great deal of junk, like a hard drive riddled with old software programs and obsolete downloads that interrupt basic processes, like those cursed pop-ups offering you discount cosmetics from long-defunct websites when all you're trying to do is read an email.

Bottom line: the brain is fallible. It may be the seat of consciousness and the engine of all human experience, but it's also incredibly messy and disorganised despite these profound roles. You have only to look at the thing to grasp how ridiculous it is: it resembles a mutant walnut, a Lovecraftian blancmange, a decrepit boxing glove, and so on. It's undeniably impressive, but it's far from perfect, and *these imperfections influence everything humans say, do and experience.*

So rather than the brain's more haphazard properties being downplayed or just flat out ignored, they should be emphasised, celebrated even. This book covers the many things the brain does that are downright laughable and how they affect us. It also looks at some of the ways people have thought the brain works that have proved to be way off. Readers of this book should, I hope, come away with a better and reassuring understanding of why people (or they themselves) regularly do and say such weird things, as well as with the ability to legitimately raise a sceptical eyebrow when confronted with the increasing amount of brain-based neuro-nonsense in the modern world. If this book can claim to have anything as lofty as overarching themes or aims, these are they.

And my final apology is based on the fact that a former colleague of mine once told me that I'd get a book published 'when hell freezes over'. Sorry to Satan. This must be very inconvenient for you.

Dean Burnett, PhD (no, really)

1

Mind controls

How the brain regulates the body, and usually makes a mess of things

The mechanics that allow us to think and reason and contemplate didn't exist millions of years ago. The first fish to crawl onto land aeons ago wasn't racked with self-doubt, thinking, 'Why am I doing this? I can't breathe up here and I don't even have any legs, whatever they are. This is the last time I play truth-or-dare with Gary.' No; until relatively recently, the brain had a much more clear and simple purpose: keeping the body alive by any means necessary.

The primitive human brain was obviously successful because we as a species endured and are now the dominant life-form on earth. But despite our evolved complicated cognitive abilities, the original primitive brain functions didn't go away. If anything, they became more important; having language and reasoning skills doesn't really amount to much if you keep dying from simple things like forgetting to eat or wandering off cliffs.

The brain needs the body to sustain it, and the body needs the brain to control it and make it do necessary things. (They're actually far more intertwined than this description suggests, but just go with it for now.) As a result, much of the brain is dedicated to basic physiological processes, monitoring internal workings, coordinating responses to problems, cleaning up mess. Maintenance, essentially. The regions that control these fundamental aspects, the brainstem and cerebellum, are sometimes referred to as the 'reptile' brain, emphasising their primitive nature, because it's the same thing the brain was doing when we were reptiles, back in the mists of time. (Mammals were a later addition to the whole 'life-on-earth' scene.) By contrast, all the more advanced abilities we modern humans enjoy – consciousness, attention, perception, reasoning – are found in the neocortex, 'neo' meaning 'new'. The actual arrangement is far more complex than these labels suggest, but it's a useful shorthand.

So you might hope that these parts – the reptile brain and the neocortex – would work together harmoniously, or at least ignore each other. Some hope. If you've ever worked for someone who's a micromanager, you know how incredibly inefficient this arrangement can be. Having someone less experienced (but technically higher ranking) hovering over you, issuing ill-informed orders and asking dumb questions can only ever make it harder. The neocortex does this with the reptile brain all the time.

It's not all one way though. The neocortex is flexible and responsive; the reptile brain is set in its ways. We've all met people who think they know best because they're older or have been doing something for longer. Working with these people can be a nightmare, like trying to write computer programs with someone who insists on using a typewriter because 'that's how it's always been done'. The reptile brain can be like that, derailing useful things by being incredibly obstinate. This chapter looks at how the brain messes up the more basic functions of the body.

Stop the book, I want to get off! (How the brain causes motion sickness)

Modern humans spend a lot more time sitting down than ever before. Manual-labour jobs have largely been replaced by office jobs. Cars and other means of transport mean we can travel while sitting down. The Internet means it is possible to spend practically your whole life sitting down, what with telecommuting, online banking and shopping.

This has its down sides. Obscene sums are spent on ergonomically designed office chairs to make sure people don't get damaged or injured due to excessive sitting. Sitting too long on an aeroplane can even be fatal, due to deep vein thrombosis. It seems odd, but very little movement is damaging.

Because moving is important. Humans are good at it and we do it a lot, as evidenced by the fact that, as a species, we've pretty much covered the surface of the earth, and actually been to the moon. Walking two miles a day has been reported as being good for the brain, but then it's probably good for every part of the body.[1] Our skeletons have evolved to allow long periods of walking, as the arrangement and properties of our feet, legs, hips and general body layout are ideally suited to regular ambulation. But it's not just the structure of our bodies; we're seemingly 'programmed' to walk without even getting the brain involved.

There are nerve clusters in our spines that help control our locomotion without any conscious involvement.[2] These bundles of nerves are called pattern generators, and are found in the lower parts of the spinal cord in the central nervous system. These pattern generators stimulate the muscles and tendons of the legs to move in specific patterns (hence the name) to produce walking. They also receive feedback from the muscles, tendons, skin and joints – such as detecting if we're walking down a slope – so we can tweak and adjust the manner of movement to match the situation. This may explain why an unconscious person can still wander about, as we'll see in the phenomenon of sleepwalking later in this chapter.

This ability to move around easily and without thinking about it – whether fleeing dangerous environments, finding food sources, pursuing prey or outrunning predators – ensured our species's survival. The first organisms to leave the sea and colonise the land led to all air-breathing life on earth; they wouldn't have done so if they'd stayed put.

But here's the question: if moving is integral to our well-being and survival, and we've actually evolved sophisticated biological systems to ensure it happens as often and as easily as possible, why does it sometimes make us throw up? This is the phenomenon known as motion sickness or travel sickness. Sometimes, often apropos of nothing, being in transit makes us bring up our breakfast, lose our lunch, or eject some other more recent but non-alliterative meal.

It's the brain that's actually responsible for this, not the stomach or innards (despite how it may feel at the time). What possible reason could there be for our brains to conclude, in defiance of aeons of evolution, that going from A to B is a legitimate cause for vomiting? In actual fact, the brain isn't defying our evolved tendencies at all. It's the numerous systems and mechanisms we have to facilitate motion that are causing the problem. Motion sickness occurs only when you're travelling by artificial means –when you're in a vehicle. Here's why.

Humans have a sophisticated array of senses and neurological mechanisms that give rise to proprioception, the ability to sense how our body is currently arranged, and which parts are going where. Put your hand behind your back and you can still sense the hand, know where it is and what rude gesture it's making, without actually seeing it. That's proprioception.

There's also the vestibular system, found in our inner ear. It's a bunch of fluid-filled canals (meaning 'bony tubes' in this context) to detect our balance and position. There's enough space in there for fluid to move about in response to gravity, and there are neurons throughout it that can detect the location and arrangement of the fluids, letting our brain know our current position and orientation. If the fluid is at the top of the tubes, this means we're upside-down, which probably isn't ideal and should be remedied as soon as possible.

Human motion (walking, running, even crawling or hopping) produces a very specific set of signals. There's the steady up–down rocking motion inherent in bipedal walking, the general velocity and the external forces such as the movement of air passing you and your shifting internal fluids that this produces. All of these are detected by proprioception and the vestibular system.

The image hitting our eyes is one of the outside world going by. The same image could be caused either by us moving or by us staying still and the outside world going past. At the most basic level, both are valid interpretations. How does the brain know which is right? It receives the visual information, couples it with the information from the fluid system in the ear and concludes 'body is moving; this is normal', and then goes back to thinking about sex or revenge or Pokemon, whatever it is you're into. Our eyes and inner systems work together to explain what's going on.

Movement via a vehicle produces a different set of sensations. Cars don't have that signature rhythmical rocking motion that our brains associate with walking (unless your suspension is well and truly shot), and the same usually goes for planes, trains and ships. When you're being transported, you're not the one actually 'doing' the moving; you're just sitting there doing something to pass the time, such as trying to stop yourself from throwing up. Your proprioception isn't producing all those clever signals for the brain to comprehend what's going on. No signals means you're not doing anything to the reptile brain, and this is reinforced by your eyes telling it you're not moving. But you *are* actually moving, and the aforementioned fluids in your ear, responding to the forces caused by high-speed movement and acceleration, are sending signals to the brain that are saying you are travelling, and quite fast at that.

What's happening now is that the brain is getting mixed signals from a precisely calibrated motion-detection system, and it is believed that this is what causes motion sickness. Our conscious brain can handle this conflicting information quite easily, but the deeper, more fundamental subconscious systems that regulate our bodies don't really know how to deal with internal problems like this, and they've no idea what could possibly be happening to cause the malfunction. In fact, as far as the reptile brain is concerned, there's only one likely answer: poison. In nature, that's the only likely thing that can so deeply affect our inner workings and cause them to get so confused.

Poison is bad, and if the brain thinks there's poison in the body, there's only one reasonable response: get rid of it, activate the vomiting reflex, pronto. The more advanced brain regions may know better, but it takes a lot of effort to alter the actions of the fundamental regions once they're under way. They are 'set in their ways' after all, almost by definition.

The phenomenon is still not totally understood at present. Why don't we get motion sickness all the time? Why do some people never suffer from it? There may well be many external or personal factors, such as the exact nature of the vehicle in which you are travelling, or some neurological predisposition to sensitivity to certain forms of movement, that contribute to occurrence of motion sickness, but this section sums up the most popular current theory. An alternative explanation is the 'nystagmus hypothesis',[3] which argues that the inadvertent stretching of the extra-ocular muscles (the ones that hold and move the eyes) due to motion stimulates the vagus nerve (one of the main nerves that control the face and head) in weird ways, leading to motion sickness. In either case, we get motion sickness because our brain gets easily confused and has a limited number of options when it comes to fixing potential problems, like a manager who's been promoted above his or her ability level and responds with buzzwords and crying fits when asked to do anything.

Seasickness seems to hit people the hardest. On land there are many items in the landscape to look at that reveal your movements (for instance, trees going past); on a ship there's usually just the sea and things that are too far away to be of any use, so the visual system is even more likely to assert that there's no movement happening. Travelling on the sea also adds an unpredictable up–down motion that gets the ear fluids firing off even more signals to an increasingly confused brain. In Spike Milligan's war memoir *Adolf Hitler: My Part in His Downfall*, Spike was transferred to Africa by ship during World War II, and was one of the only soldiers in his squad who didn't succumb to seasickness. When asked what the best way to deal with seasickness was, his reply was simply, 'Sit under a tree.' There's no supporting research available, but I'm fairly confident this method would work to prevent airsickness too.

Room for pudding?
(The brain's complex and confusing control of diet and eating)

Food is fuel. When your body needs energy, you eat. When it doesn't, you don't. It should be so simple when you think about it, but that's exactly the problem: us big smart humans can and do *think* about it, which introduces all manner of problems and neuroses.

The brain exerts a level of control over our eating and appetite that might surprise most people.* You'd think it's all controlled by the stomach or intestines, perhaps with input from the liver or fat reserves, the places where digested matter is processed and/or stored. And indeed, they do have their part to play, but they aren't as dominant as you might think.

* It's not exactly a one-way relationship either. The brain doesn't just influence the food we eat; it seems the food we eat does (or did) have considerable influence over how our brains work.[4] There's evidence to suggest that the discovery of cooking meant humans could suddenly obtain a great deal more nourishment from food. Perhaps an early human tripped and dropped his mammoth steak into the communal campfire. The determined primitive maybe got a stick and hooked his steak out, only to find it was suddenly *more* palatable and appetising. Raw food being cooked means it's easier to eat and digest. The long and dense molecules in it are broken down or denatured, allowing our teeth, stomachs and

Take the stomach; most people say they feel 'full' when they've eaten enough. This is the first major space in the body in which consumed food ends up. The stomach expands as you fill it, and the nerves in the stomach send signals to the brain to suppress appetite and stop eating, which makes perfect sense. This is the mechanism exploited by those weight-loss milkshakes you drink instead of eating meals.[5] The milkshakes contain dense stuff that fills the stomach quickly, expanding it and sending the 'I'm full' messages to the brain without you having to pack it with cake and pies.

They are, however, a short-term solution. Many people report feeling hungry less than 20 minutes after drinking one of these shakes, and that's largely because the stomach expansion signals are just one small part of the diet and appetite control. They're the bottom rung of a long ladder that goes all the way up to the more complex elements of the brain. And the ladder occasionally zigzags or even goes through loops on the way up.[6]

It's not just the stomach nerves that influences our appetite; there are also hormones that play a role. Leptin is a hormone, secreted by fat cells, that decreases appetite. Ghrelin is released by the stomach, and increases appetite. If you have more fat stores, you secrete more appetite-suppressing hormone; if your stomach is noticing a persistent emptiness, it secretes hormone to increase appetite. Simple, right? Unfortunately, no. People may have increased levels of these

---

intestines to get better nourishment from our food. This seemingly led to a rapid expansion in brain development. The human brain is an incredibly demanding organ when it comes to bodily resources, but cooking food allowed us to meet its needs. Enhanced brain development meant we got smarter, and invented better ways of hunting, and methods of farming and agriculture and so on. Food gave us bigger brains, and bigger brains gave us more food, forming a literal feedback.

hormones depending on their food requirements, but the brain can quickly grow used to them and effectively ignores them if they persist too long. One of the brain's more prominent skills is the ability to ignore anything that becomes too predictable, no matter how important it may be (this is why soldiers can still get some sleep in war zones).

Have you noticed how you always have 'room for dessert'? You might have just eaten the best part of a cow, or enough cheesy pasta to sink a gondola, but you can manage that fudge brownie or triple-scoop ice-cream sundae. Why? *How?* If your stomach is full, how is eating more even physically possible? It's largely because your brain makes an executive decision and decides that, no, you still have room. The sweetness of desserts is a palpable reward that the brain recognises and wants (see Chapter 8) so it overrules the stomach, saying, 'No room in here.' Unlike the situation with motion sickness, here the neocortex overrules the reptile brain.

Exactly why this is the case is uncertain. It may be that humans *need* quite a complex diet in order to remain in tip-top condition, so rather than just relying on our basic metabolic systems to eat whatever is available, the brain steps in and tries to regulate our diet better. And this would be fine if that was all the brain does. But it doesn't. So it isn't.

Learned associations are incredibly powerful when it comes to eating. You may be a big fan of something like, say, cake. You can be eating cake for years without any bother, then one day you eat some cake that makes you sick. Could be some of the cream in it has gone sour; it might contain an ingredient you're allergic to; or (and here's the annoying one) *it could be that something else entirely made you sick shortly after eating cake*. But, from then on, your brain has made the connection and considers cake out of bounds; if you even look at it again it can trigger the nausea response. The disgust association is a particularly powerful one, evolved to stop us eating poison or diseased things, and it can be a hard one to break. No matter that your body has consumed it dozens of times with no problem; the brain says, *No!* And there's little you can do about it.

But it doesn't have to be anything as extreme as being sick. The brain interferes with almost every food-based decision. You may have heard that the first bite is with the eye? Much of our brain, as much as 65 per cent of it, is associated with vision rather than taste.[7] While the nature and function of the connections is staggeringly varied, it does reveal that vision is clearly the go-to sensory information for the human brain. By contrast, taste is almost embarrassingly feeble, as we shall see in Chapter 5. If blindfolded while wearing nose plugs, your typical person can often mistake potato for apple.[8] Clearly, the eyes have a much greater influence over what we perceive than the tongue, so how food looks is going to influence strongly how we enjoy it, hence all the effort on presentation in the fancy eateries.

Routine can also drastically influence your eating habits. To demonstrate this, consider the phrase 'lunchtime'. When is lunchtime? Most will say between 12 p.m. and 2 p.m. Why? If food is needed for energy, why would everyone in a population, from hard physical workers like labourers and lumberjacks to sedentary people like writers and programmers, eat lunch at the same time? It's because we all agreed long ago that this was lunchtime and people rarely question it. Once you fall into this pattern, your brain quickly expects it to be maintained, and you'll get hungry *because it's time to eat*, rather than *knowing it's time to eat* because you're hungry. The brain apparently thinks logic is a precious resource to be used only sparingly.

Habits are a big part of our eating regime, and once our brain starts to expect things, our body quickly follows suit. It's all very well saying to someone who's overweight that they just need to be more disciplined and eat less, but it's not that easy. How you ended up overeating in the first place can be due to many factors, such as comfort eating. If you're sad or depressed, your brain is sending signals to the body that you're tired and exhausted. And if you're tired and exhausted, what do you need? Energy. And where do you get energy? *Food!* High-calorie food can also trigger the reward and pleasure circuits in our brains.[9] This is also why you rarely ever hear of a 'comfort salad'.

But once your brain and body adapts to a certain caloric intake, it can be very hard to reduce it. You've seen sprinters or marathon runners after a race, doubled up and gasping for breath? Do you ever consider them a glutton for oxygen? You never see anyone tell them they're lacking in discipline and are just being lazy or greedy. It's a similar effect (albeit a less healthy one) with eating, in that the body changes to expect the increased food intake, and as a result it becomes harder to stop. The exact reasons why someone ends up eating more than they need in the first place and becoming accustomed to it are impossible to determine as there are so many possibilities, but you could argue that it's an inevitability when you make endless amounts of food available to a species that has evolved to take whatever food it can get whenever it can get it.

And if you need any further proof that the brain controls

eating, consider the existence of eating disorders such as anorexia or bulimia. The brain manages to convince the body that body image is more important than food, so *it doesn't need food!* This is akin to you convincing a car that it doesn't need petrol. It's neither logical nor safe, and yet it happens worryingly regularly. Moving and eating, two basic requirements, are made needlessly complex due to our brains interfering with the process. However, eating is one of life's great pleasures, and if we were to treat it as if we were just shovelling coal into a furnace, maybe our lives would be a lot duller. Maybe the brain knows what it's doing after all.

To sleep, perchance to dream . . . or spasm, or suffocate, or sleepwalk
(The brain and the complicated properties of sleep)

Sleep involves doing literally nothing, lying down and not being conscious. How complicated could it possibly be?

Very. Sleep, the actual workings of sleep, how it happens and what's going on during it, is something people don't really think about that often. Logically, it's very hard to think about sleep while it's happening, what with the whole 'being unconscious' thing. This is a shame because it's baffled many scientists, and if more people thought about it we might be able to figure it out faster.

To clarify; we *still don't know* the purpose of sleep! It's been observed (if you adopt a fairly loose definition) in almost every other type of animal, even the simplest kinds like nematodes, a basic and common parasitic flatworm.[10] Some animals, such as jellyfish and sponges, don't show any sign of sleeping, but they don't even have brains so you can't trust them to do much of anything. But sleep, or at least some regular period of inactivity, is seen in a wide variety of radically different species. Clearly it's important, with deep evolutionary origins. Aquatic mammals have evolved methods of sleeping with only half the brain at a time because if they slept fully they'd stop swimming, sink and drown. Sleep is so important it outranks 'not drowning', and yet we don't know why.

There are many existing theories, such as healing. Rats deprived of sleep have been shown to recover much more slowly from wounds and generally don't live nearly as long as rats that get sufficient sleep.[11] An alternative theory is that sleep reduces the signal strength of weak neurological connections to make them easier to remove.[12] Another is sleep facilitates reduction of negative emotions.[13]

One of the more bizarre theories is that sleep evolved a means of preserving us from predators.[14] A lot of predators are active at night, and humans don't need 24 hours of activity to sustain themselves, so sleep provides prolonged periods where people are essentially inert, and not giving off the signs and cues that a nocturnal predator could use to find them.

Some may scoff at the cluelessness of modern scientists. Sleep is for rest, where we give our body and brain time to recover and recharge after a day's exertions. And, yes, if we've been doing something particularly exhausting, a prolonged period of inactivity is helpful for letting our systems recover and replenish/rebuild where necessary.

But if sleep is all about resting, why do we almost always sleep *for the same length of time* whether we've spent the day hauling bricks or sitting in our pyjamas watching cartoons? Surely, both activities don't require equivalent recuperation time. And metabolic activity of the body during sleep lowers by only 5 per cent to 10 per cent. This is only slightly 'relaxing' – like dropping from 50 mph to 45 mph while driving because there's smoke coming from the engine is only slightly helpful.

Exhaustion doesn't dictate our sleep patterns, which is why people seldom just fall asleep while running a marathon. Rather, the timing and duration of sleep is determined by our body's circadian rhythms, set by specific internal mechanisms. There's the pineal gland in the brain that regulates our sleep pattern via secretion of the hormone known as melatonin, which makes us relaxed and sleepy. The pineal gland responds to light levels. The retinas in our eyes detect light and send signals to the pineal gland, and the more signals it receives the less melatonin it releases (although it does still produce it at lower levels). The melatonin levels in our body rise gradually throughout the day, and increase more rapidly when the sun goes down, hence our circadian rhythms are linked to daylight hours so we're usually alert in the morning and tired at night.

This is the mechanism behind jet-lag. Travelling to another time zone means you are experiencing a completely different schedule of daylight, so you may be experiencing 11 a.m. levels of daylight when your brain thinks it's 8 p.m. Our sleep cycles are very precisely attuned, and this throwing off of our melatonin levels disrupts them. And it's harder to 'catch up' on sleep than you'd think; your brain and body are tied to the circadian rhythm, so it's difficult to force sleep at a time when it's not expected (although not impossible). A few days of the new light schedule and the rhythms are effectively reset.

You might wonder, if our sleep cycle is so sensitive to light levels, why doesn't artificial light affect them? Well, it does. People's sleep patterns now have apparently changed wildly in the last few centuries since artificial light became commonplace, and sleep patterns differ depending on culture.[15] Cultures with less access to artificial light or different daylight patterns (for example, at higher latitudes) have sleep patterns that have adapted to their circumstances.

Our core body temperature also changes according to similar rhythms, varying between 37°C and 36°C (which is a big variation for a mammal). It's highest in the afternoon, then drops as evening approaches. At midway between the highest and lowest points is when we typically go to bed, so we're asleep when it's at its lowest, which may explain the human tendency to insulate ourselves with blankets while we sleep; we're colder then than when we're awake.

To challenge further the assumption that sleep is all about rest and conserving energy, sleep has been observed in hibernating animals.[16] That is, in animals that are *already unconscious*. Hibernation isn't the same as sleep; the metabolism and body temperature drops much lower; it lasts longer; it's closer to a coma really. But hibernating animals regularly enter a sleep state, so they *use more energy in order to fall asleep!* This idea that sleep is about rest is clearly not the whole story.

This is especially true of the brain, which demonstrates complicated behaviours during sleep. Briefly, there are currently four stages of sleep: rapid-eye-movement sleep (REM) and three non-rapid-eye-movement (NREM) stages (NREM Stage 1, NREM Stage 2 and NREM Stage 3, in a rare example of neuroscientists keeping things simple for the lay person). The three NREM stages are differentiated by the type of activity the brain displays during each.

Often the different areas in the brain synchronise their patterns of activity, resulting in what you might call 'brainwaves'. If other people's brains start synchronising too, this is called a 'Mexican brainwave'.* There are several types of brainwaves, and each NREM stage has specific ones that occur.

In NREM Stage 1 the brain displays largely 'alpha' waves; NREM Stage 2 has weird patterns called 'spindles', and NREM Stage 3 is predominately 'delta' waves. There is a gradual reduction in brain activity as we progress through the sleep stages, and the further you progress the harder you are to wake up. During NREM Stage 3 sleep – 'deep' sleep – an individual is far less responsive to external stimulus such as someone yelling, 'Wake up! The house is on fire!', than at Stage 1. But the brain never shuts down completely, partly because it has several roles in maintaining the sleep state, but mostly because if it did shut down completely we'd be dead.

Then we have REM sleep, where the brain is as active, if not more so, as when we're awake and alert. One interesting (or sometimes terrifying) feature of REM sleep is REM atonia. This is where the brain's ability to control movement via motor neurons is essentially switched off, leaving us unable to move. Exactly how this happens is unclear; it could be that specific neurons inhibit activity in the motor cortex, or the sensitivity of the motor control areas is reduced, making it much harder to trigger movements. Regardless of how it occurs, it does.

And that's a good thing, too. REM sleep is when dreaming occurs, so if the motor system was left fully operational

* This is a joke. For now.

people would be physically acting out what they're doing in their dreams. If you can remember anything you've done in your dreams, you can probably see why this would be something you'd want to avoid. Thrashing and flailing while asleep and unaware of your surroundings is potentially very dangerous, for you and any unfortunate person sleeping nearby. Of course, the brain isn't 100 per cent reliable, so there are cases of REM behavioural disorders, where the motor paralysis isn't effective and people do in fact act out their dreams. And it's as hazardous as I've suggested, resulting in phenomena such as sleepwalking, which we'll get to shortly.

There are also more subtle glitches which will probably be more familiar to the everyday person. There's the hypnic jerk, where you twitch suddenly and unexpectedly while falling asleep. It feels as if you're falling suddenly, resulting in spasm while in bed. This occurs more in children and gradually declines as we age. The occurrence of hypnic jerks has been associated with anxiety, stress, sleep disorders and so on, but overall they seem to be largely random. Some theories state it's the brain mistaking falling asleep for 'dying', so it tries urgently to wake us up. But this makes little sense as the brain needs to be complicit in us falling asleep. Another theory is that it's an evolutionary holdover from a time when we slept in trees and sudden tilting or tipping sensations meant we were about to fall out, so the brain panics and wakes us. It could even be something else entirely. The reason it occurs more in children is likely to be due to the brain still being in the developing stages, where connections are still being wired up and processes and functions are being ironed out. In many ways we never truly get rid of *all* the glitches and kinks in such complicated systems as those used by our brains, so hypnic jerks persist into adulthood. Overall it's just a bit odd, if essentially harmless.[17]

What's also mostly harmless, but doesn't feel like it, is sleep paralysis. For some reason, the brain sometimes forgets to switch the motor system back on when we regain consciousness. Exactly how and why this happens hasn't been confirmed, but the dominant theories link it to disruption of the neat organisation of the sleep states. Each stage of sleep is regulated by different types of neuronal activity, and these are regulated by different sets of neurons. It can happen that the differing activity doesn't alter smoothly, so the neuronal signals that reactivate the motor system are too weak, or the ones that shut it down are too strong or last too long, and as such we regain consciousness without regaining motor control. Whatever it is that shuts down movement during REM sleep is still in place when we become fully alert, so we're unable to move.[18] This typically doesn't last long as once we wake up the rest of the brain activity resumes normal conscious levels and overrides the sleep system signals, but while it does it can be terrifying.

This terror is not unrelated either; the helplessness and vulnerability of sleep paralysis triggers a powerful fear response. This mechanism of this will be discussed in the next section, but it can be intense enough to trigger hallucinations of danger, giving rise to feelings of another presence in the room, and this is believed to be the root cause of alien-abduction fantasies, and the legend of the succubus. Most people who experience sleep paralysis do so only briefly and very rarely, but in some it can be a chronic and persistent concern. It has been linked to depression and similar disorders, suggesting some underlying issue with brain processing.

Even more complex, but likely to be related to sleep

paralysis, is the occurrence of sleepwalking. This has also been traced to the system that shuts off motor control of the brain during sleep, except now it's the reverse – that the system isn't powerful or coordinated enough. Sleepwalking is more common in children, leading scientists to theorise sleepwalking is due to the motor inhibition system being not yet fully developed. Some studies point to hints of underdevelopment in the central nervous system as a likely cause (or at least contributing factor).[19] Sleepwalking has been observed as heritable and more common in certain families, suggesting that a genetic component might underlie this central nervous system immaturity. But sleepwalking can also occur in adults under the influence of stress, alcohol, medications and so forth, any or all of which might also affect this motor inhibition system. Some scientists argue that sleepwalking is a variation or expression of epilepsy, which of course is the result of uncontrolled or chaotic brain activity, which seems logical in this instance. However it's expressed, it's invariably alarming when the brain gets the sleep and motor control functions mixed up.

But this wouldn't be an issue if the brain wasn't so active during sleep to begin with. So why is it? What's it doing in there?

The highly active REM sleep stage has a number of possible roles. One of the main ones involves memory. One persistent theory is that during REM sleep the brain is reinforcing and organising and maintaining our memories. Old memories are connected to new memories; new memories are activated to help reinforce them and make them more accessible; very old memories are stimulated to make sure the connections to them aren't lost entirely, and so on. This process takes place during sleep, possibly because there is no external information coming in to the brain to confuse or complicate matters. You never come across roads being resurfaced while cars are still going over them, and the same logic applies here.

But the activation and maintenance of the memories causes them to be effectively 'relived'. Very old experiences and more recent imaginings are all thrown into the mix together. There's no specific order or logical structure to the sequence of experiences this results in, hence dreams are invariably so other-worldly and bizarre. It's also theorised that the frontal regions of the brain responsible for attention and logic are trying to impose some sort of rationale on this ramshackle sequences of events, which explains why we still feel as if dreams are real while they're happening and the impossible occurrences don't strike as unusual at the time.

Despite the wild and unpredictable nature of dreams, certain dreams can be recurring, and these are usually associated with some issue or problem. Indeed, if there's a certain thing in your life stressing you out (like a deadline for finishing a book you've agreed to write) then you're going to think about this a lot. As a result, you'll have a lot of new memories about it that need to be organised, so will occur more in dreams, so it crops up more often and you end up regularly dreaming about burning down a publisher's office.

Another theory about REM sleep is that it's especially important for small children as it aids neurological development, going beyond just memories and shoring up and reinforcing all the connections in the brain. This would help explain why babies and the very young have to sleep a lot more than adults (often more than half the day) and spend a great deal longer in REM sleep (about 80 per cent of total sleep time as opposed to about 20 per cent in adults). Adults retain REM sleep but at lower levels to keep the brain efficient.

Yet another theory is that sleep is essential to clear out the waste products of the brain. The ongoing complex cellular processes of the brain produce a wide variety of by-products that need to be cleared away, and studies have shown that this occurs at a higher rate during sleep, so it could be that sleep for the brain is the equivalent of a restaurant closing down to clear up between lunchtime and evening openings; it's just as busy, but doing different things.

Whatever the true reason for it, sleep is essential for normal brain functioning. People deprived of sleep, particularly of REM sleep, quickly show a serious decline in cognitive focus, attention and problem-solving skills, an increase in stress levels, lower moods, irritability, and a drop in all-round task performance; the nuclear disasters of Chernobyl and Three Mile Island have been linked to overworked and exhausted engineers, so has the *Challenger* shuttle disaster, and let's not go into the long-term consequences of decisions made by sleep-deprived doctors on their third successive twelve-hour shift in two days.[20] If you go too long without sleep, your brain starts initiating 'micro sleeps', where you grab snatches of sleep for minutes or even seconds at a time. But we've evolved to expect and utilise long periods of unconsciousness, and we can't really make do with small crumbs here and there. Even if we do manage to persevere with all the cognitive problems a lack of sleep causes, it's associated with impaired immune systems, obesity, stress and heart problems.

So if you happen to nod off while reading this book, it's not boring, it's medicinal.

It's either an old dressing gown or a bloodthirsty axe murderer
(The brain and the fight-or-flight response)

As living, breathing humans, our survival depends on our biological requirements – sleeping, eating, moving – being met. But these aren't the only things essential to our existence. There are plenty of dangers lurking in the wider world, just waiting for the opportunity to snuff us out. Luckily, millions of years of evolution have equipped us with a sophisticated and reliable system of defensive measures in order to respond to any potential threat, coordinated with admirable speed and efficiency by our marvellous brains. We even have an emotion dedicated to recognising and focusing on threats: fear. One down side of this is that our brains have an inherent 'better safe than sorry' approach that means we regularly experience fear in situations where it's not really warranted.

Most people can relate to this. Maybe you were lying awake in a dark bedroom when the shadows on the walls started looking less like the branches of the dead tree outside and more like the outstretched skeletal arms of some hideous monster. Then you see the hooded figure by the door.

It's clearly the axe murderer your friend told you about. So, obviously, you collapse into a terrified panic. The axe murderer doesn't move though. He can't. Because he's not an axe murderer, he's a dressing-gown. The one you hung up on the bedroom door earlier.

It makes no logical sense, so why on earth do we have such powerful fear reactions to things that are clearly utterly harmless? Our brains, however, aren't convinced of this harmlessness. We could all live in sterilised bubbles with every sharp edge smoothed down, but as far as the brain is concerned death could come leaping out of the nearest bush at any point. To our brains, daily life is like tightrope-walking over a vast pit full of furious honey badgers and broken glass; one wrong move and you'll end up as a gruesome mess in temporary but exquisite pain.

Such a tendency is understandable. Humans evolved in a hostile, wild environment with dangers at every turn. Those humans who developed a healthy paranoia and jumped at shadows (that genuinely may have had teeth) survived long enough to pass on their genes. As a result, when presented with any conceivable threat or danger, the modern human has a suite of (mostly unconscious) response mechanisms providing a reflex that enable them to deal better with said threat, and this reflex is still very much alive and kicking (as are humans, thanks to it). This reflex is the fight-or-flight response, which is a great name as it concisely but accurately describes its function. When presented with a threat, people can either fight it or run away.

The fight-or-flight response starts in the brain, as you'd expect. Information from the senses reaches the brain and enters the thalamus, which is basically a central hub for the brain. If the brain were a city, the thalamus would be like the main station where everything arrives before being sent to where it needs to be.[21] The thalamus connects to both the advanced conscious parts of the brain in the cortex and the more primitive 'reptile' regions in the midbrain and brain-stem. It's an important area.

Sometimes the sensory information that reaches the thalamus

is worrying. It might be unfamiliar, or familiar but worrying in context. If you're lost in the woods and you hear a growl, that's unfamiliar. If you're home alone and you hear footsteps upstairs, that's familiar, but in a bad way. In either case, the sensory information reporting this is tagged 'This isn't good.' In the cortex, where it's processed further, the more analytical part of the brain looks at the information and wonders 'Is this something to worry about?' while checking the memory to see if anything similar has occurred before. If there's not enough information to determine that whatever we're experiencing is safe, it can trigger the fight-or-flight response.

However, as well as the cortex, the sensory information is relayed to the amygdala, the part of the brain responsible for strong emotional processing, and fear in particular. The amygdala doesn't do subtlety; it senses something might be amiss and initiates a red alert straight away, a response far faster than the more complex analysis in the cortex could ever hope to be. This is why a scary sensation, like a balloon popping unexpectedly, produces a fear response almost instantly, before you can process it enough to realise it's harmless.[22]

The hypothalamus is then signalled. This is the region right under the thalamus (hence the name), and is largely responsible for 'making things happen' in the body. To extend my earlier metaphor, if the thalamus is the station, the hypothalamus is the taxi rank outside it, taking important things into the city where they get stuff done. One of the roles of the hypothalamus is triggering the fight-or-flight response. It does this by getting the sympathetic nervous system to put the body effectively at 'battle stations'.

At this point you may be wondering, 'What's the sympathetic nervous system?' Good question.

The nervous system, the network of nerves and neurons spread throughout the body, allows the brain to control the body and the body to communicate with and influence the brain. The central nervous system – the brain and the spinal cord – is where the big decisions are made, and as such these areas are protected by a sturdy layer of bone (the skull and the spinal column). But many major nerves branch out from these structures, dividing and spreading further until they innervate (the actual term for supplying organs and tissue with nerves) the rest of the body. These far-reaching nerves and branches, outside the brain and spinal cord, are referred to as the peripheral nervous system.

The peripheral nervous system has two components. There's the somatic nervous system, also known as the voluntary nervous system, which links the brain to our musculoskeletal system to allow conscious movement. There's also the autonomic nervous system, which controls all the unconscious processes that keep us functioning, so is largely linked to internal organs.

But, just to make it more complicated, the autonomic nervous system also has two components: the sympathetic and parasympathetic nervous systems. The parasympathetic nervous system is responsible for maintaining the more calm processes of the body, such as gradual digestion after meals or regulating the expulsion of waste. If someone were to make a sitcom starring the different parts of the human body, the parasympathetic nervous system would be the laidback character, telling people to 'chill out' while rarely getting off the couch.

In contrast, the sympathetic nervous system is incredibly highly strung. It would be the twitchy paranoid one, constantly wrapping itself in tin foil and ranting about the CIA to anyone who'll listen. The sympathetic nervous system is often labelled the fight-or-flight system because it's what causes the various responses the body employs to deal with threats. The sympathetic nervous system dilates our pupils, to ensure more light enters our eyes so we can better spot dangers. It increases the heart rate while shunting blood away from peripheral areas and non-essential organs and systems (including digestion and salivation – hence the dry mouth when we're scared) and towards the muscles, to ensure that we have as much energy as possible for running or fighting (and feel quite tense as a result).

The sympathetic system and parasympathetic systems are constantly active and usually balance each other and ensure normal functioning of our bodily systems. But in times of emergency, the sympathetic nervous system takes over and adapts the body for fighting or (metaphorical) flying. The fight-or-flight response triggers the adrenal medulla (just above the kidneys) as well, meaning our bodies are flooded with adrenalin, which produces many more of the familiar responses to a threat: tension, butterflies in the stomach, rapid breathing for oxygenation, even relaxing of the bowels (you don't want to be carrying unnecessary 'weight' while running for your life).

Our awareness is also turned up, making us extra sensitive to potential dangers, reducing our ability to concentrate on any minor issues we were dealing with before the scary thing happened. This is the result of both the brain being alert to danger anyway and by the adrenalin suddenly hitting it, enhancing some forms of activity and limiting others.[23]

The brain's emotional processing also steps up a gear,[24] largely because the amygdala is involved. If we're dealing with a threat, we need to be motivated to take it on or get away from it asap, so we rapidly become intensely fearful or angry, providing further focus and ensuring we don't waste time with tedious 'reasoning'.

When faced with a potential threat, both brain and body rapidly shift to a state of enhanced awareness and physical readiness to deal with it. But the problem with this is the 'potential' aspect. The fight-or-flight response kicks in *before* we know whether it's actually needed.

Again, this makes logical sense; the primitive human who runs from something that *might* be a tiger was more likely to survive and reproduce than the one who said, 'Let's just wait so we can be sure.' The first human arrives back at the tribe intact, whereas the second is the tiger's breakfast.

This is a useful survival strategy in the wild but for the modern human it's quite disruptive. The fight-or-flight response involves many real and demanding physical processes, and it takes time for the effects of these to wear off. The adrenalin surge alone takes a while to leave the bloodstream, so having our whole bodies enter combat mode whenever a balloon pops unexpectedly is rather inconvenient.[25] We can experience all the tension and build-up required for a fight-or-flight response, only to realise quickly that it's not required. But we still have tense muscles and a rapid heartbeat and so on, and not relieving this with a frantic sprint or wrestling session with an intruder can cause cramps, knots in muscles, trembling and many other unpleasant consequences as the tension becomes too much.

There's also the increased emotional sensation. Someone primed to be terrified or angry can't just switch it off in an instant, so it often ends up being directed at less deserving targets. Tell an incredibly tense person to 'relax' and see what happens.

The demanding physical aspect of the fight-or-flight response is only part of the issue. The brain being so attuned to seek out and focus on danger and threats is increasingly problematic. Firstly, the brain can take account of the present situation and become more alert to danger. If we're in a darkened bedroom, the brain is aware that we can't see as much, so is attuned for any suspicious noise, and we know it should be quiet at night, so any noises that *do* occur get far more attention and are more likely to trigger our alarm systems. Also, our brain's complexity means humans now have the ability to anticipate, rationalise and imagine, meaning we can now be scared of things that haven't happened or aren't there such as the axe-murderer dressing-gown.

Chapter 3 is dedicated to the weird ways in which the brain uses and processes fear in our daily lives. When not overseeing (and often disrupting) the fundamental processes we need to keep ourselves alive, our conscious brains are exceptionally good at thinking up ways in which we might come to harm. And it doesn't even have to be physical harm; it can be intangible things such as embarrassment or sadness, things that are physically harmless but that we still really want to avoid, so the mere possibility is enough to set off our fight-or-flight response.

2

The gift of memory (keep the receipt)

The human memory system, and its strange features

The word 'memory' is often heard these days, but in the technological sense. Computer 'memory' is an everyday concept that we all understand – a storage space for information. Phone memory, iPod memory, even a USB flash drive is referred to as a 'memory stick'. There's not much simpler than a stick. So you could forgive people for thinking that computer memory and human memory are roughly the same in terms of how they work. Information goes in, the brain records it, and you access it when you need it. Right?

Wrong. Data and info are put into the memory of a computer, where they remain until needed, at which point they are retrieved, barring some technical fault, in exactly the same state in which they were first stored. So far, so logical.

But imagine a computer that decided some information in its memory was more important than other information, for reasons that were never made clear. Or a computer that filed information in a manner that didn't make any logical sense, meaning you had to search through random folders and drives trying to find the most basic data. Or a computer that kept opening your more personal and embarrassing files, like the ones containing all your erotic Care Bears fan fiction, without being asked, and at random times. Or a computer that decided it didn't really like the information you've stored, so altered it for you to suit its preferences.

Imagine a computer that did *all* these things, *all the time*. Such a device would be flung out of your office window less than half an hour after being switched on, for an urgent and terminal meeting with the concrete car park three storeys below.

But your brain does *all these things* with your memory, and all the time. Whereas with computers you can buy a newer model or take a malfunctioning one back to the shop and scream at the salesperson who recommended it, we're basically stuck with our brain. You can't even turn it off and on again to reboot the system (sleep doesn't count, as we saw earlier).

This is just one example of why 'the brain is like a computer' is something you should say to many modern neuroscientists, if you enjoy watching people twitch due to barely suppressed frustration. This is because it's a very simplistic and misleading comparison, and the memory system is a perfect illustration of this. This chapter looks at some of the more baffling and intriguing properties of the brain's memory system. I would have described them as 'memorable', but there's no way to guarantee that, given how convoluted the memory system can be.

Why did I just come in here?
(The divide between long-term and short-term memory)

We've all done it, at some time or other. You're doing something in one room, when it suddenly occurs to you that you need to go to a different room to get something. Along the way, something distracts you – a tune on the radio, someone saying something amusing as you pass, or suddenly figuring out a plot twist in a TV show that's been bugging you for months. Whatever it is, you reach your destination and suddenly have no idea why you decided to go there. It's frustrating, it's annoying, it's time-wasting; it's one of the many quirks thrown up by the surprisingly complex way the brain processes memory.

The most familiar division in human memory for most people is that between short-term memory and long-term memory. These differ considerably, but are still interdependent. Both are appropriately named; short-term memories last about a minute max., whereas long-term memories can and do stay with you your whole life. Anyone referring to something they recall from a day or even a few hours ago as 'short-term memory' is incorrect; that's long-term memory.

Short-term memory doesn't last long, but it deals with actual conscious manipulation of information; the things we're currently thinking about, in essence. We can think about them because they're in our short-term memory; that's what it's for. Long-term memory provides copious data to aid our thinking, but it's short-term memory that actually does the thinking. (For this reason, some neuroscientists prefer to say 'working' memory, which is essentially short-term memory plus a few extra processes, as we'll see later.)

It will surprise many to find that the capacity of short-term memory is so small. Current research suggests the average short-term memory can hold a maximum of four 'items' at any one time.[1] If someone is given a list of words to remember, they should be able to remember only four words. This is based on numerous experiments where people were made to recall words or items from a previously shown list and on average could recall only four with any certainty. For many years, the capacity was believed to be seven, plus or minus two. This was labelled as the 'magic number' or 'Miller's law' as it was derived from 1950s experiments by George Miller.[2] However, refinements and reassessment of legitimate recall and experimental methods have since provided data to show the actual capacity is more like four items.

The use of the vague term 'item' isn't just poor research on my part (well, not *just* that); what actually counts as an item in short-term memory varies considerably. Humans have developed strategies to get around limited short-term-memory capacity and maximise available storage space. One of these is a process called 'chunking', where a person groups things together into a single item, or 'chunk', to better utilise their short-term memory capacity.[3] If you were asked to remember the words 'smells', 'mum', 'cheese', 'of', and 'your', that would be five items. However, if you asked to remember the phrase 'Your mum smells of cheese', that would be one item, and a possible fight with the experimenter.

In contrast, we don't know the upper limit of the long-term-memory capacity as nobody has lived long enough to fill it, but it's obscenely capacious. So why is short-term memory so restricted? Partly because it's constantly in use. We're experiencing and thinking about things at every waking moment (and some sleeping ones), which means information is coming and going at an alarmingly speedy rate. This isn't somewhere that's going to lend itself well to long-term storage, which requires stability and order – it would be like leaving all your boxes and files in the entrance of a busy airport.

Another factor is that short-term memories don't have a 'physical' basis; short-term memories are stored in specific patterns of activity in neurons. To clarify: 'neuron' is the official name for brain cells, or 'nerve' cells, and they are the basis for the whole nervous system. Each one is essentially a very small biological processor, capable of receiving and generating information in the form of electrical activity across the cell membranes that give it structure, as well as forming complex connections with other neurons. So short-term memory is based on neuronal activity in the dedicated regions responsible, such as the dorsolateral prefrontal cortex in the frontal lobe.[4] We know from brain scanning that a lot of the more sophisticated, 'thinking', stuff goes on in the frontal lobe.

Storing information in patterns of neuronal activity is a bit tricky. It's a bit like writing a shopping list in the foam on your cappuccino; it's technically possible, as the foam will retain the shapes of words for a few moments, but it's not got any longevity, and hence can't be used for storage in any practical sense. Short-term memory is for rapid processing and manipulation, and with the constant influx of information anything unimportant is ignored, and quickly overwritten or allowed to fade away.

This isn't a foolproof system. Quite often, important stuff gets bumped out of short-term memory before it can be properly dealt with, which can lead to the 'Why did I just come in here?' scenario. Also, short-term memory can become overtaxed, unable to focus on anything specific while being bombarded with new information and demands. Ever seen someone amid some hubbub (such as a children's party, or a frantic work meeting) with everyone clamouring to be heard, suddenly declare, 'I can't think with all this going on!'?

They're speaking very literally; their short-term memory isn't equipped to cope with that workload.

Obvious question: if the short-term memory where we do our thinking has such a small capacity, how the hell do we get anything done? Why aren't we all sitting around trying and failing to count the fingers on one hand? Luckily, short-term memory is linked to long-term memory, which takes a lot of pressure off.

Take a professional translator; someone listening to long detailed speech in one language and translating it into another, in real time. Surely this is more than short-term memory can cope with? Actually, it isn't. If you were asking someone to translate a language in real time *while actually learning the language*, then, yes, that would be a big ask. But for the translator the words and structure of the languages are already stored in long-term memory (the brain even has regions specifically dedicated to language, like Broca's and Wernicke's areas, as we'll see later). Short-term memory has to deal with the order of the words and the meaning of the sentences, but this is something it can do, especially with practice. And this short-term/long-term interaction is the same for everyone; you don't have to learn what a sandwich is every time you want a sandwich, but you can forget that you wanted one by the time you get to the kitchen.

There are several ways information can end up as long-term memory. At a conscious level, we can ensure that short-term memories end up as long-term memories by rehearsing the relevant information, such as a phone number of someone important. We repeat it to ourselves to ensure we can remember it. This is necessary because, rather than patterns of brief activity like short-term memories, long-term memories are based on new connections between neurons, supported by synapses, formation of which can be spurred on by doing something like repeating specific things you want to remember.

Neurons conduct signals, known as 'action potentials', along their length in order to transmit information from the body to the brain or vice versa, like electricity along a surprisingly squidgy cable. Typically, many neurons in a chain make up a nerve and conduct signals from one point to another, so signals have to travel from one neuron to the next in order to get anywhere. The link between two neurons (or possibly more) is a synapse. It's not a direct physical connection; it's actually a very narrow gap between the end of one neuron and the beginning of another (although many neurons have multiple beginning and end points, just to keep things confusing). When an action potential arrives at a synapse, the first neuron in the chain squirts chemicals known as neurotransmitters into the synapse. These travel across the synapse and interact with the membrane of the other neuron via receptors. Once a neurotransmitter interacts with a receptor, it induces another action potential in this neuron, which travels along to the next synapse, and so on. There are many different types of neurotransmitter, as we'll see later; they underpin practically all the activity of the brain, and each type of neurotransmitter has specific roles and functions. They also have specific receptors that recognise and interact with them, much like security doors that will open only if presented with the right key, password, fingerprint or retinal scan.

Synapses are believed to be where the real information is 'held' in the brain; just as a certain sequence of 1s and 0s on a hard drive represents a specific file, so a specific collection of synapses in a specific place represents a memory, which we experience when these synapses are activated. So these synapses are the physical form for specific memories. Just like certain patterns of ink on paper become, when you look at them, words that make sense in a language you recognise, similarly, when a specific synapse (or several synapses) becomes active, the brain interprets this as a memory.

This creation of new long-term memories by forming these synapses is called 'encoding'; the process where the memory is actually stored in the brain.

Encoding is something the brain can do fairly quickly, but not immediately, hence short-term memory relies on less permanent but more rapid patterns of activity to store information. It doesn't form new synapses; it just triggers a bunch of essentially multipurpose ones. Rehearsing something in short-term memory keeps it 'active' long enough to give the long-term memory time to encode it.

But this 'rehearsing something until I remember it' method isn't the only way we remember things, and we clearly don't do it for *everything* we can remember. We don't need to. There's strong evidence to suggest that nearly everything we experience is stored in the long-term memory in some form.

All of the information from our senses and the associated emotional and cognitive aspects is relayed to the hippocampus in the temporal lobe. The hippocampus is a highly active brain region that is constantly combining the never-ending streams of sensory information into 'individual' memories. According to a great wealth of experimental evidence, the hippocampus is the place that the actual encoding happens. People with a damaged hippocampus can't seem to encode new memories; those who have constantly to learn and remember new information have surprisingly large hippocampi (like taxi drivers having enlarged hippocampal regions that process spatial memory and navigation, as we'll see later), suggesting greater dependence and activity. Some experiments have even 'tagged' newly formed memories (a complex process involving injecting detectable versions of proteins used in neuronal formation) and found that they are concentrated at the hippocampus.[5] This isn't even including all the newer scanning experiments that can be used to investigate hippocampal activity in real time.

New memories are laid down by the hippocampus and slowly move out into the cortex as new memories form 'behind' them, gradually nudging them along. This gradual reinforcing and shoring up of encoded memories is known as 'consolidation'. So the short-term-memory approach of repeating something until it's remembered isn't *essential* for making new long-term memories, but it is often crucial for making sure that *a specific arrangement of information* is encoded.

Say it's a phone number. This is just a sequence of numbers that are already in the long-term memory. Why would it need to encode them again? By repeating the phone number, it flags up that this particular *sequence* of numbers is important and requires a dedicated memory to be retained long term. The repetition is the short-term memory equivalent of taking a bit of information, sticking on a label marked '*Urgent!*' then sending it to the filing team.

So, if the long-term memory remembers everything, how do we still end up forgetting things? Good question.

The general consensus is that forgotten long-term memories are still technically there in the brain, barring some trauma in which they're physically destroyed (at which point being unable to remember a friend's birthday will not seem so important). But long-term memories have to go through three stages in order to be useful: they need to be made (encoded); they need to be effectively stored (in the hippocampus and then the cortex); and they need to be retrieved. If you can't retrieve a memory, it's as good as not being there at all. It's like when you can't find your gloves; you've still *got* gloves, they still exist, but you've got cold hands anyway.

Some memories are easily retrieved because they are more salient (more prominent, relevant, intense). For example, memories for something with a great degree of emotional attachment, such as your wedding day or first kiss or that time you got two bags of crisps out of the vending machine when you only paid for one, are usually very easily recalled. As well as the event itself, there's also all the emotion and thoughts and sensations going on at the same time. All of these create more and more links in the brain to this specific memory, which means the aforementioned consolidation process attaches a lot more importance to it and adds more links to it, making it much easier to retrieve. In contrast, memories with minimal or no important associations (for instance, the 473rd uneventful commute to work) get the bare minimum of consolidation, so they're a lot harder to retrieve.

The brain even uses this as something of a survival strategy – albeit a distressing one. Victims of traumatic events often end up suffering from 'flashbulb' memories, where the memory of the car accident or gruesome crime is vivid and it keeps recurring long after the event (see Chapter 8). The sensations at the time of the trauma were so intense, with the brain and body flooded with adrenalin causing heightened senses and awareness, that the memory lodges powerfully and remains raw and visceral. It's as if the brain took stock of the awful things happening and said, 'This right here, this is awful; do *not* forget this, we *do not* want to have to go through this again.' The trouble is, the memory can be so vivid it becomes disruptive.

But no memory is formed in isolation, so even in more mundane scenarios the context in which the memory was acquired can also be used as a 'trigger' to help retrieve it, as some bizarre studies have revealed.

In one example, scientists got two groups of subjects to learn some information. One group learned it in a standard room; the other group learned it while underwater, wearing full scuba suits.[6] They were later tested on the information they were told to learn, either in the same situation or the alternative one. Those who studied and were tested in the same situation performed significantly better than those who studied and were tested in different ones. Those who studied underwater and did the test underwater got much better scores than those who studied underwater but did the test in a normal room.

Being underwater was nothing to do with the information being learned, but it was the *context* in which the information was learned, and this is a big help in accessing memory. Much of the memory for where information is learned involves the context at the time, so putting someone in the same context essentially 'activates' part of the memory, making it easier to retrieve it, like revealing several letters in a game of hangman.

At this point, it's important to point out that memories for things that happen to us are not the only types of memories. These are called episodic memories, or 'autobiographical' memories, which should be self-explanatory. But we also have 'semantic' memories, which are for information essentially without the context: you remember light travels faster than sound, but not the specific physics lesson where you learned this. Remembering that the capital of France is Paris is a semantic memory, remembering the time you were sick off the Eiffel Tower is an episodic memory.

And these are the long-term memories we're consciously aware of. There's a whole swathe of long-term memories that we *don't need to be aware of* like abilities we have without thinking about it, such as driving a car or riding a bike. These things are called 'procedural' memories, and we won't go into them any further because you'll start thinking about them, and that might make it harder to use them.

In summary, short-term memory is fast, manipulative and fleeting, whereas long-term memory is persistent, enduring and capacious. This is why a funny thing that happened while in school can be something you remember for ever, and yet still decide to go into a room but, if distracted even slightly, forget why by the time you get there.

Hey, it's . . . you! From . . . the thing . . . that time

(The mechanisms of why we remember faces before names)

'You know that girl you went to school with?'

'Can you narrow it down?'

'You know, the tall girl. Dark blond hair but I think she was dyeing it, between you and me. She used to live in the street next to us before her parents divorced and her mother moved into the flat that the Jones family lived in before they moved to Australia. Her sister was friends with your cousin before she got pregnant with that boy from town, bit of a scandal that was. Always wore a red coat but it didn't really suit her. You know who I mean?'

'What's her name?'

'No idea.'

I've had countless conversations like this, with my mother, gran or other family members. Clearly, there's nothing wrong with their memory or grasp of detail; they can provide personal data about someone that would put a Wikipedia page to shame. But so many people say they struggle with names, even when they're looking directly at the person whose name they're trying to recall. I've done this myself. It makes for a very awkward wedding ceremony.

Why does this happen? Why can we recognise someone's face but not their name? Surely both are equally valid ways of identifying someone? We need to delve a bit deeper into how human memory works to grasp what's really going on.

Firstly, faces are very informative. Expressions, eye contact, mouth movements, these are all fundamental ways humans communicate.[7] Facial features also reveal a lot about a person: eye colour, hair colour, bone structure, teeth arrangement; all things that can be used to recognise a person. So much so that the human brain has seemingly evolved several features to aid and enhance facial recognition and processing, such as pattern recognition and a general predisposition to pick out faces in random images, as we'll see in Chapter 5.

Compared to all this, what has someone's name got to offer? Potentially some clues as to their background or cultural origin, but in general it's just a couple of words, a sequence of arbitrary syllables, a brief series of noises that you're informed belong to a specific face. But so what?

As we have seen, for a random piece of conscious information to go from short-term memory to long-term memory, it usually has to be repeated and rehearsed. However, you can sometimes skip this step, particularly if the information is attached to something deeply important or stimulating, meaning an episodic memory is formed. If you meet someone and they're the most beautiful person you've ever seen and you fall instantly in love, you'd be whispering the object of your affection's name to yourself for weeks.

This doesn't usually happen when you meet someone (thankfully), so if you wish to learn someone's name, the only guaranteed way to remember it is to rehearse it while it's still in your short-term memory. The trouble is, this approach takes time and uses mental resources. And as we saw from the 'Why did I just come in here?' example, something you're thinking about can be easily overwritten or replaced by the next thing you encounter and have to process. When you first meet someone, it's extremely rare for them to tell you their name and nothing else. You're invariably going to be involved in a conversation about where you're from, what you do for work, hobbies, what they arrested you for, that sort of thing. Social etiquette insists we exchange pleasantries on first meeting (even if we're not really interested), but every pleasantry we engage in with a person increases the odds of the person's name being pushed out of short-term memory before we can encode it.

Most people know dozens of names and don't find it takes considerable effort each time you need to learn a new one.

This is because your memory associates the name you hear with the person you're interacting with, so a connection is formed in your brain between person and name. As you extend your interaction, more and more connections with the person and their name are formed, so conscious rehearsing isn't needed; it happens at a more subconscious level due to your prolonged experience of engaging with the person.

The brain has many strategies for making the most of short-term memory, and one of these is that if you are provided with a lot of details in one go, the brain's memory systems tend to emphasise the first thing you hear and the last thing you hear (known as the 'primacy effect' and 'recency effect', respectively),[8] so a person's name will probably get more weight in general introductions if it's the first thing you hear (and it usually is).

There's more. One difference between short- and long-term memory not discussed so far is that they both have different overall preferences for the *type* of information they process. Short-term memory is largely *aural*, focusing on processing information in the form of words and specific sounds. This is why you have an internal monologue, and think using sentences and language, rather than a series of images like a film. Someone's name is an example of aural information; you hear the words, and think of it in terms of the sounds that form them.

In contrast to this, the long-term memory also relies heavily on vision and semantic qualities (the *meaning* of words, rather than the sounds that form them).[9] So a rich visual stimulus, like, say, someone's face, is more likely to be remembered long term than some random aural stimulus, like an unfamiliar name.

In a purely objective sense, a person's face and name are,

by and large, unrelated. You might hear people say, 'You look like a Martin' (on learning someone's name is Martin), but in truth it's borderline impossible to predict accurately a name just by looking at a face – unless that name is tattooed on his or her forehead (a striking visual feature that is very hard to forget).

Let's say that both someone's name and face have been successfully stored in the long-term memory. Great, well done. But that's only half the battle; now you need to access this information when needed. And that, unfortunately, can prove difficult.

The brain is a terrifyingly complex tangle of connections and links, like a ball of Christmas-tree lights the size of the known universe. Long-term memories are made up of these connections, these synapses. A single neuron can have tens of thousands of synapses with other neurons, and the brain has many billions of neurons, but these synapses mean there is a link between a specific memory and the more 'executive' areas (the bits that do all the rationalisation and decision-making) such as the frontal cortex that requires the information in the memory. These links are what allows the thinking parts of your brain to 'get at' memories, so to speak.

The more connections a specific memory has, and the 'stronger' (more active) the synapse is, the easier it is to access, in the same way that it's easier to travel to somewhere with multiple roads and transport links than to an abandoned barn in the middle of a wilderness. The name and face of your long-term partner, for example, is going to occur in a great deal of memories, so it will always be at the forefront of your mind. Other people aren't going to get this treatment (unless your relationships are rather more atypical), so remembering their names is going to be harder.

But if the brain has already stored someone's face and name, why do we still end up remembering one and not the other? This is because the brain has something of a two-tier memory system at work when it comes to retrieving memories, and this gives rise to a common yet infuriating sensation: recognising someone, but not being able to remember how or why, or what their name is. This happens because the brain differentiates between familiarity and recall.[10] To clarify, familiarity (or recognition) is when you encounter someone or something and you know you've done so before. But beyond that, you've got nothing; all you can say is this person/thing is already in your memories. Recall is when you can access the original memory of how and why you know this person; recognition is just flagging up the fact that the memory exists.

The brain has several ways and means to trigger a memory, but you don't need to 'activate' a memory to know it's there. You know when you try to save a file onto your computer and it says, 'This file already exists'? It's a bit like that. All you know is that the information is there; you can't get at it yet.

You can see how such a system would be advantageous; it means you don't have to dedicate too much precious brain power to figuring out if you've encountered something before. And, in the harsh reality of the natural world, anything that's familiar is something that didn't kill you, so you can concentrate on newer things that might. It makes evolutionary sense for the brain to work this way. Given that a face provides more information than a name, faces are more likely to be 'familiar'.

But this doesn't mean it's not intensely annoying for us modern humans, who regularly have to make small talk with people we're certain we know but can't actually recall right now. That's the part most people can relate to, the point where recognition turns to full-on recall. Some scientists describe it as a 'recall threshold',[11] where something becomes increasingly familiar, until it reaches a crucial point and the original memory is activated. The desired memory has several other memories linked to it, and these are being triggered and cause a sort of peripheral or low-level stimulation of the target memory, like a darkened house being lit by a neighbour's firework display. But the target memory won't actually activate until it is stimulated above a specific level, or threshold.

You've heard the phrase 'it all came flooding back', or you recognise the sensation of a quiz question being 'on the tip of your tongue' before it suddenly occurs to you? That's what's happening here. The memory that caused all this recognition has now received enough stimulation and is finally activated, the neighbour's fireworks have woken those living in the house and they've turned all the lights on, so all the associated information is now available. Your memory is officially jogged, the tip of your tongue can resume its normal duties of tasting things rather than providing an unlikely storage space for trivia.

Overall, faces are more memorable than names because they're more 'tangible', whereas remembering someone's name is more likely to require full recall than simple recognition. I hope this information means that you'll understand that if we ever meet for a second time and I don't remember your name, I'm not being rude.

Actually, in terms of social etiquette, I probably *am* being rude. But now at least you know why.

A glass of wine to refresh your memory (How alcohol can actually help you remember things)

People like alcohol. So much so that alcohol-related issues are an ongoing problem for many populations. These issues can be so widespread and constant that dealing with them ends up costing billions.[12] So why is something so damaging also so popular?

Probably because alcohol is fun. Aside from causing a dopamine release in the areas of your brain dealing with reward and pleasure (see Chapter 8), thus causing that weird euphoric buzz that social drinkers enjoy so much. There's also social convention built up around alcohol; it's almost a mandatory element of celebration, bonding and just general recreation. Because of this, you can see why the more detrimental effects of alcohol are regularly overlooked. Sure, hangovers are bad, but comparing and laughing about the severity of respective hangovers is yet another way of bonding with friends. And the ridiculous ways in which people behave when drunk would be deeply alarming in some contexts (in a school, perhaps, at 10 a.m.) but when everyone does it, it's just fun, right? A necessary relief from the seriousness and conformity demanded of us by modern society. So, yes, the negative aspects of alcohol are considered a price worth paying by those who enjoy it.

One of these negative aspects is memory loss. Alcohol and memory loss go hand in unsteady hand. It's a comedy staple in sitcoms, stand-up and even personal anecdotes, usually involving someone waking up after a drunken night and finding himself in an unexpected situation, surrounded by traffic cones, unfamiliar garments, snoring strangers, irate swans and other things that wouldn't be in a person's bedroom under normal circumstances.

So how then can alcohol possibly actually *help* your memory, as the title of this bit suggests? Well, it's necessary to go over why alcohol affects our brain's memory system in the first place. After all, we ingest countless different chemicals and substances every time we eat anything, why don't they cause us to slur our words or pick fights with lamp-posts?

It's due to the chemical properties of alcohol. The body and brain have several levels of defence to stop potentially harmful substances entering our systems (stomach acids, complex intestinal linings, dedicated barriers to keep things out of the brain . . .) but alcohol (specifically ethanol, the type we drink) dissolves in water and is small enough to pass through all these defences, so the alcohol we drink ends up spread throughout our bodily systems via the bloodstream. And when it builds up in the brain, several bags of spanners are thrown into some very important workings.

Alcohol is a depressant.[13] Not because it makes you feel dreadful and depressed the next morning (although, good lord, it does), but because it actually depresses activity in the nerves of the brain; it reduces their activity like someone lowering the volume on a stereo. But why would this make people behave in *more* ridiculous ways? If brain activity is reduced, shouldn't drunk people just sit there quietly and dribble?

Yes, some drunk people do precisely this, but remember that the countless processes the human brain is carrying out every waking moment require not just making things happen, but *preventing* things from happening. The brain controls pretty much everything we do, but we can't do everything all at once, so much of the brain is dedicated to inhibition and stopping activation of certain brain areas. Think of the way traffic is controlled in a large city; it is a complex job, relying on 'stop' signs or red traffic lights to some degree. Without them the city would grind to a messy halt in a matter of minutes. Similarly, the brain has countless areas that provide important and essential functions but *only when needed*. For example, the part of your brain that moves your leg is very important, but not when you're trying to sit in a meeting, so you need another part of the brain to say, 'Not now, mate', to the leg-controlling part.

Under the influence of alcohol, the red traffic lights are dimmed or switched off in the brain regions that normally keep giddiness, euphoria and anger in check or suppressed. Alcohol also shuts down the areas responsible for speech clarity or walking coordination.[14]

It is worth noting that our simpler, fundamental systems, controlling things such as heart rate, are deeply entrenched and robust, whereas the newer, more sophisticated processes are more easily disrupted or damaged by alcohol. There are similar parallels in modern technology; you could drop a 1980s Walkman down a flight of stairs and it might still work, but tap a smartphone on the corner of a table and you end up with a hefty repair bill. Sophistication results in vulnerability, it seems.

So with the brain and alcohol, 'higher' functions are the first to go. Things like social restraint, embarrassment and the little voices in our head that say, 'This probably isn't a good idea.' Alcohol silences these pretty quickly. When you're drunk you're more likely to say what's on your mind or take a crazy risk just to get a laugh, such as agreeing to write an entire book about the brain.[15]

The last things to be disrupted by alcohol (and it has to be a lot to get to this point) are the basic physiological processes, such as heart rate and breathing. If you're so drunk you get into this state, you'll probably lack sufficient brain function to be capable of being worried, but you really *really* should be.[16]

Between these two extremes, there's the memory system, which is technically both fundamental and complex. Alcohol seems to have a particular tendency to disrupt the hippocampus, the main region for memory formation and encoding. It can also limit your short-term memory, but it's the long-term memory disruption via the hippocampus that causes the worrying gaps when you wake up the next day. It's not a complete shutdown of course; memories are usually still being formed, but less efficiently and more haphazardly.[17]

Interesting aside: for most people, drinking enough to block memory formation completely (alcoholic blackouts) would mean they're so intoxicated they can barely speak or stand. Alcoholics, however, are different. They've been drinking a lot for a long time, so much so that their bodies and brains have actually adapted to deal with, and even require, a regular alcohol intake, so they can remain upright and coherent (more or less) despite consuming way more alcohol than your average person could withstand (see Chapter 8).

However, the alcohol they've consumed still has an effect on the memory system, and if there's enough sloshing around in their heads it can cause a full 'shutdown' of memory formation *while they're still talking and behaving normally* thanks to their tolerance. They don't show any outward signs of problems, but ten minutes later, they've no memory of what they've been saying or doing. It's as though they stepped away from the controls of a video game and someone else took over; it looked the same to anyone watching the game, but the original player has no idea what's been happening while they were in the toilet.[18]

Yes, alcohol disrupts the memory system. But, in very specific circumstances, it can actually *help* recall. This is the phenomenon known as state-specific recall.

We've covered already how the external context can help you recall a memory; you're better able to recall it if you are in the same environment where the memory was acquired. But, and here's the clever bit, this also applies to the *internal* context, or 'state', hence state-dependent recall.[19] To put it simply, substances such as alcohol or stimulants or anything that alters brain activity bring about a specific neurological state. When the brain is suddenly having to deal with a disruptive substance washing around everywhere, this does not go unnoticed, any more than you wouldn't notice that your bedroom was suddenly full of smoke.

This can also apply to mood; if you learn something while in a bad mood, you're more likely to recall it later if you're in a bad mood again. It's a massive oversimplification to describe moods and mood disorders as 'chemical imbalances' in the brain (despite many who do just that) but the overall levels of chemical and electrochemical activity that result in and from a specific mood is something the brain can recognise, and does. Thus, the context *inside* your head is potentially just as useful as the one *outside* your head when it comes to triggering memories.

Alcohol does disrupt memories, but only after a certain point; it's perfectly possible to have the pleasant buzz of a few beers or glasses of wine and still remember everything the next day. But if you were to be told some interesting gossip or useful information after a couple of glasses of wine, your brain would encode your slightly intoxicated state as part of the memory, so would be better able to retrieve this memory if you were to have another couple of glasses of wine (on a different night, not right after the first two). In this scenario, a glass of wine can indeed improve your memory.

Please don't take this as a scientific endorsement for drinking heavily when studying for exams or tests. Turning up drunk for a test will be problematic enough to cancel out any minor memory advantages this provides you with, especially if it's a driving test.

But there is still some hope for desperate students: caffeine affects the brain and produces a specific internal state that can help trigger memories, and a lot of students pull caffeine-fuelled all-nighters when cramming for exams, so if you attend the exams similarly stimulated by excessive caffeine then it could well help with remembering some of the more important details from your notes.

It's not exactly irrefutable evidence, but I did once (unknowingly) employ this tactic at university, where I stayed up all night revising for an exam I was particularly worried about. A lot of coffee kept me going and I indulged in an extra-large mug right before the exam, to ensure I stayed conscious throughout. I ended up getting 73 per cent on the exam, one of the highest marks in my year.

I wouldn't recommend this approach though. Yes, I got a good mark, but I also desperately needed the toilet the whole time, called the examiner 'Dad' when I asked for more paper, and on the way home got into a furious row. With a pigeon.

Of course I remember it, it was my idea!
(The ego-bias of our memory systems)

Thus far, we've covered how the brain processes memory, and how it isn't exactly straightforward/efficient/consistent. Actually, there are numerous ways in which the brain's memory system leaves a lot to be desired, but at least you end up with access to reliable, accurate information, safely stored in your head for future use.

It would be lovely if that was true, wouldn't it? Sadly, the words 'reliable' and 'accurate' can rarely be applied to the workings of the brain, particularly for memory. The memories retrieved by the brain are sometimes comparable to a hairball coughed up by a cat, the product of a lot of alarming internal mangling.

Rather than a static record of information or events like pages in a book, our memories are regularly tweaked and modified to suit whatever the brain interprets as our needs (however wrong that may be). Surprisingly, memory is quite plastic (meaning flexible, malleable, not rigid) and can be altered, suppressed or misattributed in numerous ways. This is known as a memory bias. And memory bias is often driven by ego.

Obviously, some people have huge egos. They can be very memorable themselves, if just for the ways they inspire average people to fantasise many elaborate ways of killing them. But even though most people don't have a dreadful ego, they do still have an ego, which influences the nature and detail of the memories they recall. Why?

The tone of this book thus far has referred to 'the brain'

as if it's a separate self-contained entity, an approach used by most books or articles about the brain, and one that makes logical sense. If you want to provide a scientific analysis of something, then it's necessary to be as objective and rational as possible, and treat the brain as just another organ, like the heart or liver.

But it's not. Your brain is *you*. And here the subject matter edges over into the philosophical areas. Are we as individuals really just the product of a mass of neurons firing off sparks, or are we more than the sum of our parts? Does the mind really arise from the brain, or is it in fact some separate entity, intrinsically linked to it but not exactly 'the same'? What does this mean for free will and our ability to strive for higher goals? These are questions that thinkers have grappled with ever since it was figured out that our consciousness resides in the brain. (This seems obvious now, but for many centuries it was believed the heart was the seat of our minds and the brain had more mundane functions such as cooling or filtering blood. Echoes of this time still persist in our language; for example, 'Do what your heart tells you.'[20])

These are discussions for elsewhere, but suffice it to say that scientific understanding and evidence strongly imply that our sense of self and all that goes with it (memory, language, emotion, perception, and so on) is supported by processes in our brain. Everything you are is a feature of your brain, and as such much of what your brain does is dedicated to making you look and feel as good as possible, like an obsequious lackey to a popular celebrity, who prevents her hearing any criticism or negative publicity for fear of upsetting her. And one of the ways it can do this is by modifying your memories to make you feel better about yourself.

There are numerous memory biases or flaws, many of which aren't noticeably egotistical in nature. However, a surprising number appear to be largely egotistical, especially the one simply called the egocentric bias, where our memories are tweaked or modified by the brain to present events in a manner that makes us look better.[21] For example, if recalling an occasion where they were part of a group decision, people tend to remember that they were more influential and integral to the final decision than they in fact were.

One of the earliest reports of this stems from the Watergate scandal, where a whistleblower told investigators all about the plans and discussions that he had taken part in that lead to the political conspiracy and cover-up. However, later listening to the recordings of these meetings, an accurate record of the discussions, revealed John Dean got the overall 'gist' of what happened, but many of his claims were alarmingly inaccurate. The main problem was that he'd described himself as an influential key figure in the planning, but the tapes revealed he was a bit player at most. He hadn't set out to lie, just to boost his own ego; his memory was 'altered' to conform to his sense of identity and self-importance.[22]

It doesn't have to be government-toppling corruption though; it can be minor things such as believing you performed better at sports than you genuinely did, or recalling you caught a trout when it was in fact a minnow. It's important to note that when this happens it's not an example of someone lying or exaggerating to impress people; it often happens with memories *even if we're not telling anyone about them*. That last bit is key: we genuinely believe our memory's version of events to be accurate and fair. The modifications and tweaks made to give a more flattering portrayal of ourselves is, more often than not, entirely unconscious.

There are other memory biases that can be attributed to ego. There's choice-supportive bias, when you have to choose one of several options, and you remember it as being the best of all available options, even if it wasn't at the time.[23] Each option could be practically identical in terms of merit and potential outcome, but the brain alters your memory to downplay the rejected ones and big-up the option you went with, making you feel you chose wisely, even if it was totally random.

There's the self-generation effect, where you're better at recalling things that you've said than at recalling things other people have said.[24] You can never be sure how accurate or authentic someone else is being, but you believe *you* are when you say something, and seeing as it's your memory that amounts to the same thing.

More alarming is the own-race bias, where people struggle to recall and identify people from races other than their own.[25] Ego isn't exactly subtle and thoughtful, and it may be expressed in crude ways such as prioritising or emphasising people of the same or similar racial background over those who aren't, as yours is the 'best' one. You may not think this at all, but your subconscious isn't always so sophisticated.

You may have heard the saying, 'Hindsight is 20–20', usually used to dismiss someone claiming prior knowledge of an event after it's happened. It's generally assumed that the person is exaggerating or lying, because they didn't use this prior knowledge when it would actually have been useful. For example: 'If you were so certain that Barry had been drinking, why did you let him drive you to the airport?'

While it is no doubt true that some people do exaggerate

their awareness in this manner to seem smarter and better informed, there is actually such a thing as the hindsight bias in memory, where we genuinely remember past events as being predictable even though we couldn't have hoped to predict them at the time.[26] Again, this isn't some self-aggrandising fabrication, our memories genuinely do seem to support this notion. The brain alters memories to boost our ego, making us feel as if we were better informed and in control.

How about the fading-affect bias,[27] where emotional memories for negative events fade more quickly than positive ones. The memories themselves may remain intact, but the emotional component of them can fade with time, and it seems that, in general, unpleasant emotions fade faster than nice ones. The brain clearly likes it if nice things happen to you, but doesn't dwell on the 'alternative' stuff.

These are just some of the biases that could be seen as demonstrations of ego overriding accuracy. It's just something your brain does all the time. But *why*?* Surely an accurate memory of events would be far more useful than some self-serving distortion?

Well, yes and no. Only some biases have this apparent connection to ego, whereas others have the opposite. Some people demonstrate things like 'persistence', which in this case is when memories of a traumatic event keep recurring, despite the individual's lack of desire to think about them.[28] This is quite a common phenomenon, and doesn't need to be something especially damaging or disturbing. You might be

* Exactly *how* it does this is another matter altogether. It's not really established yet, and the details involving conscious influence over memory encoding and retrieval, self-oriented filtering of perception and numerous other relevant processes that may play a role probably warrant a book all of their own.

wandering along the road on your way somewhere, casually thinking about nothing in particular, and your brain suddenly says, 'Remember when you asked that girl out at the school party and she laughed in your face in front of everyone and you ran away but collided with a table and landed in the cakes?' Suddenly you're racked with shame and embarrassment thanks to a twenty-year-old memory, apropos of nothing. Other biases, like childhood amnesia or context dependence, suggest limitations or inaccuracies arising from the way the memory system works, rather than anything ego-based.

It's also important to remember that the changes caused by these memory biases are (usually) quite limited, rather than major alterations. You may remember doing better in a job interview than you actually did, but you won't remember getting the job if that didn't happen. The ego bias of the brain isn't so powerful as to create different realities; it just tweaks and adjusts recall of events, it doesn't create new ones.

But why would it do this at all? Firstly, human beings need to make a lot of decisions, and this is a lot easier if they have at least some degree of confidence when making them. The brain constructs a model of how the world works in order to navigate it, and it needs to be confident that this is accurate (see Chapter 8, the section on 'delusions', for more about this). If you had to weigh up every possible outcome for every choice you have to make, it would be extremely time consuming. This can be avoided if you have confidence in yourself and your abilities to make the right choice.

Secondly, *all* our memories are formed from a personal, subjective viewpoint. The only perspective and interpretation we have when making judgements is our own, and as a result this could lead to our memories prioritising when it was 'right' more than when it wasn't, to the extent that our judgement is protected and reinforced in memory even when it's not strictly correct.

On top of this, a sense of self-worth and achievement seems to be integral to normal functioning for humans (see Chapter 7). When people lose their sense of self-worth – for example, if they are experiencing clinical depression – it can be genuinely debilitating. But even when functioning normally, the brain is prone to worrying and dwelling on negative outcomes; like when you can't stop thinking about what *might* have happened following an important event like a job interview, even though it didn't happen – a process known as counterfactual thinking.[29] A degree of self-confidence and ego, even if artificially produced by manipulated memories, is important for normal functioning.

Some may find this quite alarming, the idea that your memories aren't reliable because of your ego. And if it applies to everyone, can you really trust what anyone says? Maybe everyone is remembering things wrongly due to subconscious self-flattery? Luckily, there's probably no need to panic; many things still get done properly and efficiently, so what ego biases there are seem to be relatively harmless overall. But still, it might be wise to retain an element of scepticism whenever hearing someone make self-aggrandising claims.

For example, in this section, I've tried to impress you by explaining memory and ego are linked. But what if I've just remembered things that supported my notion and forgotten the rest? I claimed the self-generation effect, where people remember things they've said better than things other people have said, was due to ego. But an alternative explanation is that the things you say involve your brain to a much greater extent. You've got to think of the thing to say, process it, go through the physical motions required to speak it, listen back to it, judge for reactions, so *of course* you'd remember it more.

The choice-supportive bias, where we remember our choice as being the 'best' one: an example of ego, or the brain's way of preventing us from dwelling on possibilities that did not and cannot occur? This is something humans do often, taking up a lot of valuable energy, often for no appreciable gain.

How about the cross-race effect, where people struggle to recall people's features if they're of a race not their own? Some dark side of egotistical preference, or the result of being raised among people of your own race, meaning your brain has had a lot more practice differentiating between people who are racially similar to you?

There are alternative explanations for all the biases mentioned above, other than ego. So is this whole section just the result of my own raging ego? No, not really. There is a lot of evidence to support the conclusion that egocentric bias is a genuine phenomenon, such as studies revealing that people are far more willing and able to criticise their actions from many years ago than they are more recent actions, most likely because the recent actions are a much closer portrayal of how they are now, and this is far too close to self-criticism, so is suppressed or overlooked.[30] People even show tendencies to criticise 'past' selves and praise 'present' selves even when there's been no real improvement or change in the matter in question ('I didn't learn to drive when I was a teenager because I was too lazy, but I haven't learned now because I'm too busy'). This criticism of a past self may seem to contradict egocentric memory bias, but it works to emphasise how much the present self has improved and grown and so should be proud.

The brain regularly edits memories to make them more flattering, whatever the rationale for doing so, and these edits and tweaks can become self-sustaining. If we remember and/or describe an event in a way that slightly emphasises our role in it (we caught the biggest fish on a fishing trip, rather than the third biggest), the existing memory is then effectively 'updated' with this new modification (the modification is arguably a new event, but is strongly linked to the existing memory, so the brain has to reconcile this somehow). And this happens again the next time it's recalled. And the next, and so on. It's one of those things that happens without you knowing or realising, and the brain is so complex that there are often several different explanations for the same phenomenon, all occurring simultaneously, all of which are equally valid.

The upside of this is, even if you don't quite understand what's been written about here, you'll probably remember that you did, so it all ends up the same regardless. Good work.

Where am I? . . . Who am I?
(When and how the memory system can go wrong)

In this chapter, we've covered some of the more impressive and outlandish properties of the brain's memory system, but all of these have assumed that the memory is working normally (for want of a better term). But what if things go wrong? What can happen to disrupt the brain's memory systems? We've seen that ego can distort your memory, but that it rarely if ever distorts so severely it actually creates new memories for things that didn't actually happen. This was an attempt to reassure you. Now let's undo that by pointing out that I didn't say it *never* happens.

Take 'false memories'. False memories can be very dangerous, especially if they're a false memory of something awful. There have been reports of arguably well-intentioned psychologists and psychiatrists trying to uncover repressed memories in patients who have seemingly ended up creating (supposedly by accident) the terrible memories they're trying to 'uncover' in the first place. This is the psychological equivalent of poisoning the water supply.

The most worrying thing is that you don't need to be suffering from psychological issues to have false memories created in your head; it can happen to virtually anyone. It might seem a bit ridiculous that someone can implant false memories in our brains by just talking to us, but neurologically it's not that far-fetched. Language is seemingly fundamental to our way of thinking, and we base much of our world view on what other people think of and tell us (see Chapter 7).

Much of the research on false memories is focused on eyewitness testimonies.[31] In important legal cases, innocent lives could be altered for ever by witnesses misremembering a single detail, or remembering something that didn't happen.

Eyewitness accounts are valuable in court but that's one of the worst places to obtain them. It's often a very tense and intimidating atmosphere and the people testifying are made fully aware of the seriousness of the situation, promising to 'tell the truth, the whole truth and nothing but the truth, so help me God'. Promising a judge you won't lie and invoking the supreme creator of the universe to back you up? These aren't exactly casual circumstances, and probably will cause considerable stress and distraction.

People tend to be very suggestive to those they recognise as authority figures, and one persistent finding is that when people are being quizzed about their memory, the nature of the question can have a major influence on what is remembered. The best-known name connected to this phenomenon is Professor Elizabeth Loftus, who has done extensive research into the subject.[32] She herself regularly cites the worrying cases of individuals who have had extremely traumatic memories 'implanted' (presumably accidentally) by questionable and untested therapeutic methods. A particularly famous case involves Nadine Cool, a woman who sought therapy for a traumatic experience in the 1980s and ended up with detailed memories of being part of a murderous satanic cult. This never happened though, and she ended up successfully suing the therapist for millions of dollars.[33]

Professor Loftus's research details several studies where people are shown videos of car accidents or similar occurrences and then asked questions about what was observed. It's been persistently found (in these and other studies) that the structure of the questions asked directly influences what an individual can remember.[34] Such an occurrence is especially relevant for eyewitness testimonies.

In particular conditions, such as the individual being anxious and the question coming from someone with authority (say, the lawyer in a court room), specific wording can 'create' a memory. For example, if the lawyer asks, 'Was the defendant in the vicinity of the cheese shop at the time of the great cheddar robbery?', then the witness can answer yes or no, according to what he or she remembers. But if the lawyer asks, 'Where in the cheese shop was the defendant at the time of the great cheddar robbery?', this question asserts that the defendant *was definitely there*. The witness may not remember seeing the defendant, but the question, stated as a fact from a higher-status person, causes the brain to doubt its own records, and actually adjust them to conform to the new 'facts' presented by this 'reliable' source. The witness can end up saying something like, 'I think he was stood next to the gorgonzola', and mean it, even though he or she witnessed no such thing at the time. That something so fundamental to our society should have such a glaring vulnerability is disconcerting. I was once asked to testify in a court that all the witnesses for the prosecution could just be demonstrating false memories. I didn't do it, as I was worried I could inadvertently destroy the whole justice system.

We can see just how easy it is disrupt the memory *when it's functioning normally*. But what if something actually goes wrong with the brain mechanisms responsible for memory? There are a number of ways this can happen, none of which are particularly nice.

At the extreme end of the scale, there's serious brain damage, such as that caused by aggressive neurodegenerative conditions such as Alzheimer's disease. Alzheimer's (and other forms of dementia) is the result of widespread cell death throughout the brain, causing many symptoms, but the best known is unpredictable memory loss and disruption. The exact reason this occurs is uncertain, but one main theory at present is that it's caused by neurofibrillary tangles.[35]

Neurons are long, branching cells, and they have what are basically 'skeletons' (called cytoskeletons) made of long protein chains. These long chains are called neurofilaments, and several neurofilaments combined into one 'stronger' structure, like the strands making up a rope, is a neurofibril. These provide structural support for the cell and help transport important substances along it. But, for some reason, in some people, these neurofibrils are no longer arranged in neat sequences, but end up tangled like a garden hose left unattended for five minutes. It could be a small but crucial mutation in a relevant gene causing the proteins to unfold in unpredictable ways; it could be some other currently unknown cellular process that gets more common as we age. Whatever the cause, this tangling seriously disrupts the workings of the neuron, choking off its essential processes, eventually causing it to die. And this happens throughout the brain, affecting almost all the areas involved in memory.

However, damage to memory doesn't have to be caused by a problem that occurs at the cellular level. Stroke, a disturbance in the blood supply to the brain, is also particularly bad for memory; the hippocampus, responsible for encoding and processing all our memories at all times, is an incredibly resource-intensive neurological region, requiring an uninterrupted supply of nutrients and metabolites. Fuel, essentially. A stroke can cut off this supply, even briefly, which is a bit like pulling the battery out of a laptop. Brevity is irrelevant; the damage is done. The memory system won't be working so well from now on. Although there is some hope, in that it has to be a powerful or particularly precise stroke (blood has many ways of getting to the brain) to cause serious memory problems.[36]

There's a difference between 'unilateral' and 'bilateral' strokes. In simple terms, the brain has two hemispheres, both of which have a hippocampus; a stroke that affects both is pretty devastating, but a stroke that affects just one hemispheres is more manageable. Much has been learned about the memory system from subjects who have suffered varying memory deficits from strokes, or even weirdly precise injuries. One subject referenced in scientific studies on memory was an amnesia sufferer whose condition resulted from somehow getting a snooker cue lodged right up his nose to the point where it physically damaged his brain.[37] There's really no such thing as a 'non-contact' sport.

There have even been cases where the memory-processing parts of the brain have been removed deliberately via surgery. This is how areas of the brain responsible for memory were recognised in the first place. In the days before brain scans and other flashy technology, there was Patient HM. Patient HM suffered severe temporal-lobe epilepsy, meaning the areas of his temporal lobe were causing debilitating fits so often that it was determined that they had to be removed. So they were, successfully, and the fits stopped. Unfortunately, so did his long-term memory. From then on, Patient HM could remember only the months leading up to surgery, and no more. He could remember things that happened to him less than a minute ago, but then he'd forget them. This is how it was established that the temporal lobe is where all the memory-formation workings are in the brain.[38]

Patients with hippocampal amnesia are still studied today, and the wider-reaching functions of the hippocampus is constantly being established. For example, a recent study from 2013 suggests that hippocampal damage impairs creative thinking ability.[39] It makes sense; it must be harder to be creative if you can't retain and access interesting memories and combinations of stimuli.

Perhaps as interesting were the memory systems HM *didn't* lose. He clearly retained his short-term memory, but information in short-term memory no longer had anywhere to go, so it faded away. He could learn new motor skills and abilities such as specific drawing techniques, but every time you tested him on a specific ability, he was convinced it was the first time he'd ever attempted it, despite being quite proficient at it. Clearly, this unconscious memory was processed elsewhere by different mechanisms that had been spared.*

Soap operas would lead you to believe that 'retrograde amnesia' is the most common occurrence, meaning an inability to recall memories acquired before a trauma occurs. This is typically demonstrated by a character receiving a blow to the head (he fell and hit it in an unlikely plot device), regaining consciousness and asking, 'Where am I? Who are you people?', before slowly revealing he can't recall the past twenty years of his life.

This is far more unlikely than TV implies; the whole blow-to-the-head-and-lose-whole-life-story-and-identity thing is very rare. Individual memories are spread throughout the brain, so any injury that actually destroys them is likely to

* A lecturer once told me that one of the few things that HM did learn was where the biscuits were stored. But he never had any memory of having just eaten any biscuits, so he kept going back for more. He never gained memories, but he did gain weight. I can't confirm this; I've not found any direct reports or evidence for it. However, there is a study where Jeffrey Brunstrom and his team, at the University of Bristol, told hungry subjects they'd be fed either 500 ml or 300 ml of soup. They were then fed these amounts. But an ingenious set-up using discreet pumps meant that some subjects who were given 300 ml had their bowls stealthily refilled so they actually consumed 500 ml, whereas some given 500 ml had their bowls stealthily drained so they only ended up eating 300 ml.[40]

The interesting finding was that the actual amount consumed was irrelevant; it was the amount the subject *remembered* eating (however wrongly) that dictated when they got hungry. Those who thought they had consumed 300 ml of soup but had consumed 500 ml reported getting hungry much earlier than those who thought they had consumed 500 ml but had eaten 300 ml. Clearly, memory can overrule actual physiological signals when it comes to determining appetite, so it looks as if serious memory disruption can have a marked effect on diet.

destroy much of the whole brain as well.[41] If this happens, remembering your best friend's name probably isn't a priority. Similarly, the executive regions in the frontal lobe responsible for recollection are also pretty important for things such as decision-making, reasoning etc., so if they're disrupted then memory loss will be a relatively minor concern compared with the more pressing problems. People can and do demonstrate retrograde amnesia, but it is usually transient and memories eventually return. This doesn't make for good dramatic plots, but it's probably better for the individual.

If and when retrograde amnesia does occur, the nature of the disorder means it's very hard to study; it is difficult to assess and monitor the extent of someone's memory loss from their earlier life, because how would you know anything about this time? The patient could say, 'I think I remember going to the zoo on a bus when I was eleven', and it seems as though their memory is returning, but unless the doctor was actually on the bus with them at the time, how can anyone be sure? It could easily be a suggested or created memory. So in order to test and measure someone's memory loss from their earlier life, you'd need an accurate record of *their whole life* to measure any gaps or losses accurately, and having such a thing is rare.

The study of one type of retrograde amnesia resulting from a condition known as Wernicke-Korsakoff syndrome, typically the result of thiamine deficiency due to excessive alcoholism,[42] benefited from an individual known as 'Patient X', a sufferer who had previously written an autobiography. This enabled doctors to study the extent of his memory loss more precisely as they had a reference to go from.[43] We might see this happening more in the future, with more and more people charting their lives online via social media sites. But then, what people do online isn't always an accurate reflection of their lives. You can imagine clinical psychologists accessing an amnesia patient's Facebook profile and assuming their memories should consist of mostly laughing at funny videos of cats.

The hippocampus is easily disrupted or damaged – by physical trauma, stroke, various types of dementia. Even Herpes Simplex, the virus responsible for cold sores, can occasionally turn very aggressive and attack the hippocampus.[44] And, of course, as the hippocampus is essential for the formation of new memories, the more likely type of amnesia is anterograde: the inability to form new memories following a trauma. This is the sort of amnesia Patient HM suffered from (he died in 2008 at the age of seventy-eight). If you saw the film *Memento*, it's just like that. If you saw the film *Memento* but don't really remember it, that's not quite so helpful (but is ironic).

This is just a brief overview of the many things that can go wrong with the brain's memory processes, via injury, surgery, disease, drink, or anything else. Very specific types of amnesia can occur (for example, forgetting memory for events but not for facts) and some memory deficits have no recognisable physical cause (some amnesias are believed to be purely psychological, stemming from denial or reaction to traumatic experiences).

How can such a convoluted, confusing, inconsistent, vulnerable and fragile system be of any use at all? Simply because, most of the time, it *does* work. It's still awesome, with a capacity and adaptability that puts even the most modern supercomputers to shame. The inherent flexibility and weird organisation is something that's evolved over millions of years, so who am I to criticise? Human memory isn't perfect, but it's good enough.

3

Fear: nothing to be scared of

The many ways in which the brain makes us constantly afraid

What are you worrying about right now? Loads of things, probably.

Have you got everything you need for your child's upcoming birthday party? Is the big work project going as well as it could be? Will your gas bill be more than you can afford? When did your mother last call; is she OK? That ache in your hip hasn't gone away; are you sure it's not arthritis? That left-over mince has been in the fridge for a week; what if someone eats it and gets food poisoning? Why is my foot itching? Remember when your pants fell down in school when you were nine; what if people still think about that? Does the car seem a bit sluggish to you? What's that noise? Is it a rat? What if it has the plague? Your boss will never believe you if you call in sick with that. On and on and on and on and on and on.

As we saw in the earlier fight-or-flight section, our brain is primed to think up potential threats. One arguable down side of our sophisticated intelligence is that the term 'threat' is up for grabs. At one point in our dim evolutionary past, it focused only on actual, physical, life-endangering hazards, because the world was basically full of them, but those days are long gone. The world has changed, but our brains haven't caught up yet, and can find literally *anything* to fret about.

The extensive list above is just the smallest tip of the gargantuan neurotic iceberg created by our brains. Anything that might have a negative consequence, no matter how small or subjective, is logged as 'worth worrying about'. And sometimes even that isn't needed. Have you ever avoided walking under ladders, or thrown salt over your shoulder, or stayed indoors on Friday the 13th? You have all the signs of being superstitious – you are genuinely stressing about situations or processes *that have no real basis in reality*. As a result, you then behave in ways that can't realistically have any effect on events, just to feel safer.

Equally, we can get sucked into conspiracy theories, getting worked up and paranoid about things that are technically possible but incredibly unlikely. Or the brain can create phobias – we get distressed about something that we understand is harmless but we are massively afraid of nonetheless. At other times, the brain doesn't even bother coming up with even the most tenuous reason for being worried and just worries about literally nothing. How many times have you heard people say it's 'too quiet', or that things have been uneventful so something bad is 'due'. This sort of thing can afflict a person with chronic anxiety disorder. This is just one way in which the brain's tendency to worry can have actual physical effects on our bodies (high blood pressure, tension, trembling, weight loss/gain) and impact our general lives – in obsessing over harmless things, it actually causes us harm. Surveys by bodies including the Office for National Statistics (ONS) have reported that 1 in 10 adults in the UK will experience an anxiety-related disorder at some point in their lives,[1] and in its 2009 report 'In the Face of Fear', UK Mental Health revealed a percentage rise of 12.8 in anxiety-related conditions between 1993 and 2007.[2] That's nearly a million more UK adults who suffer from anxiety problems.

Who needs predators when we have our expanded craniums to drag us down with persistent stress?

What do four-leaf clovers and UFOs have in common?

(The connection between superstition, conspiracy theories and other bizarre beliefs)

Here's some interesting trivia for you: I'm involved in many shadowy conspiracies that are secretly controlling society. I'm in league with 'Big Pharma' to suppress all natural remedies, alternative medicine and cancer cures for the sake of profit (nothing spells 'big money' like potential consumers constantly dying). I'm part of a plot to ensure that the public never realises that the moon landings were an elaborate sham. My day job in the field of mental healthcare and psychiatry is obviously a massive racket intended to crush free thinkers and to enforce conformity. I'm also part of the great conspiracy of global scientists to promote the myths of climate change, evolution, vaccination and a spherical earth. After all, there's nobody on earth wealthier and more powerful than scientists, and they can't risk losing this exalted position by people finding out how the world really works.

You may be surprised to hear of my involvement in so many conspiracies. It certainly stunned me. I found out only by accident thanks to the rigorous work of the commenters below many of my *Guardian* articles. Amid suggestions that

I am the worst writer in all of time, space and humanity, and I really should go and do unspeakable physical acts with my mother/pets/furniture, you will find 'proof' of my nefarious and manifold conspiracy involvement.

This is apparently to be expected when you contribute things to a major media platform, but I was still shocked. Some of the conspiracy theories didn't even make sense. When I wrote a piece to defend transgender people after a particularly vicious article attacking them (not one that I wrote, I hasten to add), I was accused of being part of an anti-transgender people conspiracy (because I didn't defend them aggressively enough) and a pro-transgender people conspiracy (because I defended them at all). Not only am I involved in many conspiracies, I'm also actively opposing myself in the process.

It's common for readers, seeing any article critical of their existing views or beliefs, to immediately conclude it's the work of a sinister power hell-bent on suppression, rather than a prematurely balding bloke sitting on a sofa in Cardiff.

The arrival of the Internet and an increasingly interconnected society has been a great boon to conspiracy theories; people can more easily find 'evidence' for their theories on 9/11 or share their wild conclusions regarding the CIA and AIDS with like-minded types, without ever leaving the house.

Conspiracy theories aren't a new phenomenon,[3] so perhaps it's a quirk of the brain that means people are so willing and able to be swallowed up by paranoid imaginings? In a way, it is. But, going back to the title, what's this got to do with superstition? Declaring that UFOs are real and trying to break into Area 51 is a far cry from thinking a four-leaf clover is good luck, so what's the connection?

An ironic question, as it's the tendency to see patterns in

(often unrelated) things that links both conspiracies and superstitions. There's actually a name for the experience of seeing connections in places where there actually aren't any: apophenia.[4] For example, if you accidentally wear your underpants inside out and then later win some money on a scratch card, and from then on you only ever wear your underpants inside out when buying scratch cards, that's apophenia; there's no possible way your underwear orientation can affect the value of a scratch card, but you've seen the pattern and are going with it. Similarly, if two unrelated but high-profile figures die of natural causes or in accidents within a month of each other, that's tragic. But if you look at the two individuals and find they were both critical of a certain political body or government and conclude that they were in fact assassinated as a result, that's apophenia. At their most basic levels, any conspiracy or superstition can likely be traced back to someone constructing a meaningful connection between unrelated occurrences.

It's not just the extremely paranoid or suspicious types who are prone to this, anyone can experience it. And it's pretty easy to see how this could come about.

The brain receives a constant stream of varied information and it has to make some sense of this. The world we perceive is the end result of all the processing the brain does with it. From the retina to the visual cortex to the hippocampus to the prefrontal cortex, the brain relies on many different areas to perform several different functions all working in tandem. (Those newspaper reports about neuroscientific 'discoveries', implying that a specific function of the brain has a specific region dedicated to it and it alone, are misleading. This is only a partial explanation at best.)

Despite numerous brain regions being involved in sensing

and perceiving the world around us, there are still major limitations; it's not that the brain is underpowered, it's just that we're bombarded by exceptionally dense information at all times, only some of which has any relevance to us, and the brain has barely a fraction of a second to process it for us to use. And because of this, the brain has numerous short cuts it employs to keep on top of things (more or less).

One of the ways the brain sorts out the important information from the unimportant is by recognising and focusing on patterns. Direct examples of these can be observed in the visual system (see Chapter 5), but suffice it to say that the brain is constantly looking for links in the things we observe. This is undoubtedly a survival tactic, stemming from a time when our species faced constant danger – remember fight or flight? – and no doubt sets up a few false alarms. But what's a few false alarms if your survival is ensured?

But these false alarms are what cause problems. We end up with apophenia, and add to that the brain's fight-or-flight response and our tendency to leap to a worst-case-scenario conclusion and suddenly we have a lot on our minds. We see patterns in the world that don't exist, then attach serious significance to them on the off chance they may negatively affect us. Consider how many superstitions are based on avoiding bad luck or misfortune. You never hear about conspiracies that are intended to help people. The mysterious elite don't organise charity bake sales.

The brain also recognises patterns and tendencies based on information stored in the memory. The things we experience inform our ways of thinking, which makes sense. However, our first experiences are during childhood, and this informs much about our later lives. The first time you attempt to teach your parents how to use the latest video game is usually enough to dispel any remaining idea that they're all-knowing and omnipotent, but they can often seem like this during childhood. When we're growing up, much (if not all) of our environment is controlled; practically everything we know is told to us by adults we recognise and trust, everything that happens does so under their supervision. They are our primary reference points during the most formative years of our lives. So if your parents have superstitions, it's highly likely that you'll pick them up, without having to witness anything that would support them.[5]

Crucially, this also means that many of our earliest memories are formed in a world that is seemingly organised and controlled by powerful figures who are hard to understand (rather than a world that is just random or chaotic). Such notions can be deeply entrenched, and that belief system can be carried into adulthood. It is more comforting for some adults to believe that the world is organised according to the plans of powerful authority figures, be they wealthy tycoons, alien lizards with a penchant for human flesh, or scientists.

The previous paragraph may suggest that people who believe in conspiracy theories are insecure, immature individuals, subconsciously yearning for parental approval that was never forthcoming as they grew up. And no doubt some of them are, but then so are countless people who aren't into conspiracy theories; I'm not going to ramble on for several paragraphs about the risks of making ill-founded connections between two unrelated things and then do exactly that myself. What's been said is just a way of suggesting means by which the development of the brain may make conspiracy theories more 'plausible'.

But one prominent consequence (or it might be a cause) of our tendency to look for patterns is that the brain really doesn't handle randomness well. The brain seems to struggle with the idea that something can happen for no discernible reason other than chance. It might be yet another consequence of our brains seeking danger everywhere – if there's no real cause for an occurrence then there's nothing that can be done about it if it ends up being dangerous, and that's not tolerable. Or it might be something else entirely. Maybe the brain's opposition to anything random is just a chance mutation that proved useful. That would be a cruel irony, if nothing else.

Whatever the cause, the rejection of randomness has numerous knock-on consequences, one of which is the reflex assumption that everything that happens does so for a reason, often referred to as 'fate'. In reality, some people are just unfortunate, but that's not an acceptable explanation for the brain, so it has to find one and attach a flimsy rationale. Having a lot of bad luck? Must be that mirror you broke, which contained your soul, which is now fractured. Or maybe it's that you're being visited by mischievous fairies; they hate iron, so keep a horseshoe around, that'll keep them away.

You could argue that conspiracy theorists are convinced that sinister organisations are running the world because *that's better than the alternative*! The idea that all of human society is just bumbling along due to haphazard occurrences and luck is, in many ways, more distressing than there being a shadowy elite running things, even if it is for its own ends. Better a drunk pilot at the controls than nobody at all.

In personality studies, this concept is called the 'pronounced locus of control' and refers to the extent to which individuals believe they can control the events affecting them.[6] The bigger your locus of control, the more 'in control' you believe you are (the extent to which you really *are* in control of events is irrelevant). Exactly why some people feel more in control than others is a poorly understood area; some studies have linked an enlarged hippocampus to a greater locus of control,[7] but the stress hormone cortisol can apparently shrink the hippocampus, and people who feel less in control tend to be more easily stressed, so the hippocampus size may be a consequence rather than a cause of the locus of control.[8] The brain never makes anything easy for us.

Anyway, a greater locus of control means you may end up feeling you can influence the cause of these occurrences (a cause which doesn't actually exist, but no matter). If it's superstition, you throw salt over your shoulder or touch wood or avoid ladders and black cats, and are thus reassured that your actions have prevented catastrophe via means that defy all rational explanation.

Individuals with an even greater locus of control try to undermine the 'conspiracy' they see by spreading awareness of it, looking 'deeper' into the details (reliability of the source is rarely a concern) and pointing them out to anyone who'll listen, and declaring all those who don't to be 'mindless sheep' or some variation thereof. Superstitions tend to be more passive; people can just adhere to them and go about their day as normal. Conspiracy theories tend to involve a lot more dedication and effort. When was the last time someone tried to convince you of the hidden truth behind why rabbit's feet are lucky?

Overall, it seems the brain's love of patterns and hatred of randomness leads many people to make some pretty extreme conclusions. This wouldn't really be an issue, but the brain also makes it very hard to convince someone that their deeply held views and conclusions are wrong, no matter how much evidence you have. The superstitious and the conspiracy theorists maintain their bizarre beliefs despite everything the rational world throws at them. And it's all thanks to our idiot brains.

Or is it? Everything I've said here is based on the current understanding provided by neuroscience and psychology, but then that understanding is rather limited. The very subject matter alone is so hard to pin down. What is a superstition, in the psychological sense? What would one look like in the terms of brain activity? Is it a belief? An idea? We might have advanced to the point where we can scan for activity in the working brain, but just because we can see activity doesn't mean we understand what it represents, any more than being able to see a piano's keys means we can play Mozart.

Not that scientists haven't tried. For example, Marjaana Lindeman and colleagues performed fMRI scans of twelve self-described supernatural believers and eleven sceptics.[9] The subjects were told to imagine a critical life situation (such as imminent job loss or relationship breakdown) and were then shown 'emotionally charged pictures of lifeless objects and scenery (for example, two red cherries bound together)' – the sort of thing you'd see on motivational posters, like a spectacular mountain top, that sort of thing. Supernatural believers reported seeing hints and signs of how their personal situation would resolve in the image; if imagining a relationship breakdown, they would feel it would be all right because the two cherries bound together signified firm ties and commitment. The sceptics, as you'd expect, didn't do this.

The interesting element of this study is that viewing the

pictures activated the left inferior temporal gyrus in all subjects, a region associated with image processing. In the supernatural believers, much less activity was seen in the right inferior temporal gyrus when compared with the sceptics. This region has been associated with cognitive inhibition, meaning it modulates and reduces other cognitive processes.[10] In this case, it may be suppressing the activity that leads to forming illogical patterns and connections, which would explain why some people are quick to believe in irrational or unlikely occurrences while others require serious convincing; if the right inferior temporal gyrus is weak, the more irrational-leaning processes in the brain exert more influence.

This is far from a conclusive experiment though, for many reasons. For one, it's a very small number of subjects, but, mainly, how does one measure or determine one's 'supernatural leanings'? This isn't something covered by the metric system. Some people like to believe they're totally rational, but this itself may be an ironic self-delusion.

It's even worse studying conspiracy theories. The same rules apply, but it's harder to get willing subjects, given the subject matter. Conspiracy theorists tend to be secretive, paranoid and distrustful of recognised authorities, so if a scientist were to say to one, 'Would you like to come to our secure facility and let us experiment on you? It may involve being confined in a metal tube so we can scan your brain', the answer is unlikely to be yes. So all that's included in this section is a reasonable set of theories and assumptions based on the data we currently have available.

But then, I would say that, wouldn't I? This whole chapter could be part of the conspiracy to keep people in the dark . . .

Some people would rather wrestle a wildcat than sing karaoke
(Phobias, social anxieties and their numerous manifestations)

Karaoke is a globally popular pastime. Some people love getting up in front of (usually quite intoxicated) strangers and singing a song that they're often only vaguely familiar with, regardless of their singing ability. There haven't been experiments on this but I'd posit there is an inverse relationship between enthusiasm and ability. Consumption of alcohol is almost certainly a factor in this trend. And in these days of the televised talent contest, people can sing in front of millions of strangers rather than a small crowd of uninterested drunks.

To some of us, this is a terrifying prospect. The stuff nightmares are made of, in fact. You ask certain people if they want to get up and sing in front of a crowd and they'll react as if you've just told them they've got to juggle live grenades in the nude while all their ex-partners are watching. The colour will drain from their faces, they'll tense up, start breathing rapidly, and exhibit many other classic indicators of the fight-or-flight response. Given the choice between singing and taking part in combat, they'll happily engage in a fight to the death (unless there's an audience for that, too).

What's going on there? Whatever you think of karaoke, it's risk free, unless the crowd is made up of steroid-abusing music lovers. Sure, it can go badly; you might mangle a tune so awfully that everyone listening ends up begging for the sweet relief of death. But so what? So a few people you'll never meet again consider your singing abilities to be below par. Where's the harm in that? But as far as our brains are concerned, there *is* harm in that. Shame, embarrassment, public humiliation; these are all intense negative sensations that nobody but the most dedicated deviant actively seeks out. The mere possibility of any (or all) of these occurring is enough to put people off most things.

There are many things people are afraid of that are far more mundane than karaoke: talking on the telephone (something I myself avoid wherever possible), paying for something with a queue behind you, remembering a round of drinks, giving presentations, getting a haircut – things millions of people do every day without incident but that still fill some people with dread and panic.

These are social anxieties. Practically everyone has them to some extent, but if they get to the point where they are actually disruptive and debilitating to a person's functioning, they can be classed as a social phobia. Social phobias are the most common of several manifestations of phobias, so to understand the underlying neuroscience let's step back a bit and look at phobias in general.

A phobia is an *irrational* fear of something. If a spider lands on your hand unexpectedly and you yelp and flail a bit, people would understand; a creepy-crawly surprised you, people don't like insects touching them, so your reaction is justifiable. If a spider lands on your hand and you scream uncontrollably while knocking tables over before scrubbing your hand in bleach, burning all your clothes then refusing to leave your house for a month, then this may be considered 'irrational'. It's just a spider, after all.

An interesting thing about phobias is that people who have them are usually completely aware of how illogical they are.[11]

People with arachnophobia know, on a conscious level, that a spider no bigger than a penny poses no danger to them, but they can't help their excessive fear reaction. This is why the stock phrases used in response to someone's phobia ('It won't hurt you') are well meant but utterly pointless. Knowing that something isn't dangerous doesn't make much difference, so the fear we associate with the trigger obviously goes deeper than the conscious level, which is why phobias can be so tricky and persistent.

Phobias can be classed as specific (or 'simple') or complex. Both of these labels refer to the source of the phobia. Simple phobias apply to phobias of a certain object (for example, knives), animal (spiders, rats), situation (being in a lift) or thing (blood, vomiting). As long as the individual avoids these things, they're able to go about their business. Sometimes it's impossible to avoid the triggers completely, but they're usually transient; you might be scared of lifts, but a typical lift journey lasts seconds, unless you're Willy Wonka.

There are a variety of reasons for exactly *how* these phobias originate. At the most fundamental level, we have associative learning, attaching a specific response (such as a fear reaction) to a specific stimulus (such as a spider). Even the most neurologically uncomplicated creatures seem capable of it, such as Aplysia, aka the California sea slug, a very simple metre-long aquatic gastropod that was used in the 1970s in the earliest experiments to monitor neuronal changes occurring in learning.[12] They may be simple and have a rudimentary nervous system by human standards, but they can show associative learning and, more importantly, have massive neurons, big enough to stick electrodes in to record what's going on. Aplysia neurons can have axons (the long 'trunk' part of a neuron) up to a millimetre in diameter. This might not sound like much, but it's comparatively vast. If human neuron axons were the size of a drinking straw, Aplysia axons would be the size of the Channel Tunnel.

Big neurons wouldn't be of any use if the creatures couldn't show associative learning, which is the point here. We've hinted at this before; in the section on diet and appetite in Chapter 1, it was observed how the brain can make the cake–illness association and you feel sick just thinking about it. The same mechanism can apply to phobias and fears.

If you get warned against something (meeting strangers, electrical wiring, rats, germs), your brain is going to extrapolate all the bad things that could happen if you encounter it. Then you *do* encounter it, and your brain activates all these 'likely' scenarios, and activates the fight-or-flight response. The amygdala, responsible for encoding the fear component of memory, attaches a *danger* label to memories of the encounter. So, the next time you encounter this thing, you'll remember *danger*, and have the same reaction. When we learn to be wary of something we end up fearing it. In some people, this can end up as a phobia.

This process implies that literally anything can become the focus of a phobia, and if you've ever seen a list of existing phobias this seems to be the case. Notable examples include turophobia (fear of cheese), xanthophobia (fear of the colour yellow, which has obvious overlaps with turophobia), hippopotomonstrosesquipedaliophobia (fear of long words, because psychologists are basically evil) and phobophobia (fear of having a phobia, because the brain regularly turns to the concept of logic and says, 'Shut up, you're not my real dad!'). However, some phobias are considerably more common than others, suggesting that there are other factors at play.

We have *evolved* to fear certain things. One behavioural study taught chimps to be afraid of snakes. This is relatively straightforward task, usually involving showing them a snake and following this with an unpleasant sensation, like a mild electric shock or unpleasant food, just something they want to avoid if possible. The interesting part is that when other chimps saw them react fearfully to snakes, they quickly learned to fear snakes too, without having been trained.[13] This is often described as 'social learning'.*

Social learning and cues are incredibly powerful, and the brain's 'better safe than sorry' approach when it comes to dangers means if we see someone being afraid of something, there's a good chance we'll be afraid of it too. This is especially true during childhood, where our understanding of the world is still developing, largely via the input of others who we assume know more than we do. So if our parents have a particularly

* Social learning can explain much of this. We pick up much of what we know and how to behave from the actions of others, particularly if it's something like responding to a threat, and chimps are similar in that regard. Social phenomena are covered more extensively Chapter 7, but it can't be the whole explanation here, because the weird thing is that when the same procedure was performed with flowers instead of snakes, it was still possible to train chimps to fear them, but the other chimps rarely learned the same fear by observing them. Fear of snakes is easy to pass on; fear of flowers is not. We've evolved an inherent suspicion of potentially lethal dangers, hence fear of snakes and spiders is common.[14] By contrast, nobody fears flowers (anthophobia), unless they've got a particularly vicious type of hay fever. Less obvious evolved-fear tendencies include fear of lifts, or injections, or the dentist. Lifts cause us to be 'trapped', which can set off alarms in our brains. Injections and the dentist involve potential pain and invasions of body integrity, so cause fear responses. An evolved tendency to be wary or fearful of corpses (which could carry disease or indicate nearby dangers, as well as just being upsetting) may be behind the 'uncanny valley' effect,[15] where computer animations or robots that look *almost* human but not quite seem sinister and disturbing, whereas two eyes slapped on a sock is fine. The near-human construct lacks the subtle details and cues an actual human has, so seems more 'lifeless' than 'entertaining'.

strong phobia, there's a good chance we'll end up with it, like a particularly unsettling hand-me-down. It makes sense: if a child sees a parent, or their primary educator/teacher/provider/role model, start shrieking and flapping because they've seen a mouse, this is bound to be a vivid and unsettling experience, one that makes and impression on a young mind.

The brain's fear response means phobias are hard to get rid of. Most learned associations can be removed eventually via a process established in Pavlov's famous dogs experiment. A bell was associated with food, prompting a learned response (salivation) whenever it was heard, but if the bell was then rung repeatedly in the continued absence of food, eventually the association faded. This same procedure can used in many contexts, and is known as extinction (not to be confused with what happened to the dinosaurs).[16] The brain learns that the stimulus such as the bell isn't associated with anything and therefore doesn't require a specific response.

You'd think that phobias would be subject to a similar process, given how almost every encounter with their cause results in no harm whatsoever. But here's the tricky part: the fear response triggered by the phobia *justifies it*. In a masterpiece of circular logic, the brain decides that something is dangerous, and as a result it sets off the fight-or-flight response when it encounters it. This causes all the usual physical reactions, flooding our systems with adrenalin, making us tense and panicked and so on. The fight-or-flight response is biologically demanding and draining and often unpleasant to experience, so the brain remembers this as 'The last time I met that thing, the body went haywire, so I was right; it *is* dangerous!' and thus the phobia is reinforced, not diminished, regardless of how little actual harm the individual came to.

The nature of the phobia also plays a part. Thus far we've described the simple phobias (phobias triggered by specific things or objects, having an easily identified and avoidable source), but there are also complex ones (phobias triggered by more complicated things such as contexts or situations). Agoraphobia is a type of complex phobia, generally misunderstood as fear of open spaces. More precisely, agoraphobia is a fear of being in a situation where escape would be impossible or help would be absent.[17] Technically, this can be anywhere outside the person's home, hence severe agoraphobia prevents people from leaving the house, leading to the 'fear of open spaces' misconception.

Agoraphobia is strongly associated with panic disorder. Panic attacks can happen to anyone – the fear response overwhelms us and we can't do anything about it and we feel distressed/terrified/can't breathe/sick/head spins/trapped. The symptoms vary from person to person, and an interesting article by Lindsey Homes and Alissa Scheller for the *Huffington Post* in 2014 entitled 'This is what a panic attack feels like' collected some personal descriptions from sufferers, one of which was: 'Mine are like I can't stand up, I can't speak. All I feel is an intense amount of pain all over, like something is just squeezing me into this little ball. If it is really bad I can't breathe, I start to hyperventilate and I throw up.'

There are many others that differ considerably but seem just as bad.[18] It all boils down to the same thing; sometimes the brain just cuts out the middle man and starts inducing fear reactions in the absence of any feasible cause. Since there's no visible cause, there's literally nothing that can be done about the situation, so it quickly becomes 'overwhelming'. This is a panic disorder. Sufferers end up being terrified and alarmed in harmless scenarios, which they then associate with fear and panic, so end up being quite phobic towards them.

Exactly why this panic disorder occurs in the first case is currently unknown, but there are several compelling theories. It could be the result of previous trauma suffered by the individual, as the brain hasn't yet effectively dealt with the lasting issues caused. It might be to do with an excess or deficiency of particular neurotransmitters. A genetic component is possible, as those directly related to a panic disorder sufferer are more likely to experience it themselves.[19] There is even a theory that sufferers are prone to catastrophic thinking; taking a minor physical issue or problem and worrying about it far beyond what it is even vaguely rational.[20] It could be a combination of all these things, or something as yet undiscovered. The brain isn't short of options when it comes to unreasonable fear response.

And finally, we have social anxieties. Or, if they're so potent they become debilitating, social phobias. Social phobias are based on fear of negative reaction from other people – dreading your audience's reaction to your karaoke, for instance. We don't fear only hostility or aggression; simple disapproval is enough to stop us in our tracks. The fact that other people can be a powerful source of phobias is another example of how our brains use other humans to calibrate how we see the world and our position in it. As a result, the approval of others *matters*, often regardless of who they are. Fame is something millions of people strive for, and what is fame but the approval of strangers? We've already covered how egotistical the brain is, so maybe all famous people just crave mass approval? It's a bit sad really (unless they're a famous person who has praised this book).

Social anxieties occur when the brain's tendency to predict and worry about negative outcomes is combined with the brain's need for social acceptance and approval. Talking on the telephone means interacting without any of the usual cues present in person, so some people (like me) find it very difficult and we panic that we'll offend or bore the other individual. Paying for shopping with a large queue behind you can be nerve-racking as you're technically delaying a lot of people who stare at you while you try to use your maths skills working out the payments. These and countless similar situations allow the brain to work out ways in which you'll annoy or frustrate others, earning negative opinions and causing embarrassment. It boils down to performance anxiety; the worry about getting things wrong in front of an audience.

Some people have no issues with this, but some have the opposite problem. How this comes about has a variety of explanations, but a study by Roselind Lieb found that parenting styles are associated with likelihood of developing anxiety disorders,[21] and you can see the logic here. Overly critical parents can instil in a child a constant fear of upsetting a valuable authority figure for even minor actions, whereas overprotective parents can prevent a child from ever experiencing even minor negative consequences of actions, so when they're older and away from parental protection and something they do does cause a negative outcome, they're not used to it, so it affects them disproportionately, meaning they'll be less able to deal with it and will be way more likely to fear it happening again. Even having the dangers of strangers drummed into you constantly from an early age can enhance your eventual fear of them to beyond-appropriate levels.

People experiencing these phobias often display avoidant

behaviour, where they actively avoid getting into any scenario where the phobic reaction could come into play.[22] This may be good for peace of mind, but it's bad for doing anything about the phobia in the long run; the more it's avoided, the longer it stays potent and vivid in the brain. It's a bit like papering over a mouse hole in your wall; it looks fine to the casual observer, but you've still got a rodent problem.

The available evidence suggests social anxieties and phobias are apparently the most common type of phobias.[23] This isn't surprising given the brain's paranoid tendencies leading us to fear things that aren't dangerous, and our reliance on approval from others. Put these two together, and we can end up unreasonably fearful of others having a negative opinion of our incompetence. For proof of this, consider the fact that this is the ~~ninth~~ ~~tenth~~ ~~eleventh~~ ~~twelfth~~ twenty-eighth draft I've done of this conclusion. And, yes, I'm still sure loads of people won't like it.

Don't have nightmares . . . unless you're into that sort of thing.

(Why people like being scared and actively seek it out)

Why do so many people literally jump at the chance to risk smearing themselves over the unforgiving ground in pursuit of fleeting excitement? Think of base jumpers, bungee jumpers, parachutists. Everything we've learned so far shows the brain's drive for self-preservation and how that results in nervousness, avoidance behaviour, and so on. Yet authors such as Stephen King and Dean Koontz write books featuring fear-inducing supernatural occurrences and brutal, violent deaths of characters and they are raking it in. They have sold nearly a billion books between them. The *Saw* franchise, a showcase for the most inventive and gory ways in which humans can be prematurely killed for obscure reasons, currently numbers seven films, all of which were shown in cinemas worldwide rather than sealed in lead containers and launched into the sun. We tell each other scary stories around the campfire, we ride ghost trains, visit haunted houses, dress up as the walking dead at Halloween to extract sweets from neighbours. So how do we explain our enjoyment of these entertainments, some of which are aimed at children no less, that depend on us being scared?

Coincidentally, the thrill of fear and the gratification gained from sweets are both likely to be dependent on the same brain region. This is the mesolimbic pathway, often known as the mesolimbic reward pathway or the mesolimbic dopaminergic pathway, because it is responsible for the brain's sensation of reward, and it uses dopamine neurons to do it. It is one of several circuits and pathways that mediate reward, but it is largely acknowledged as being the most 'central' one. And this is what makes it important for the 'people enjoying fear' phenomenon.

This pathway is composed of the ventral tegumental area (VTA) and nucleus accumbens (NAc).[24] These are very dense collections of circuits and neural relays deep in the brain, with numerous connections and links to the more sophisticated regions including the hippocampus and the frontal lobes, and the more primitive regions such as the brainstem, so it's a very influential part of the brain.

The VTA is the component that detects a stimulus and

determines whether it was positive or negative, something to be encouraged or avoided. It then signals its decision to the NAc, which causes the appropriate response to be experienced. So if you eat a tasty snack, the VTA registers this as a good thing, tells the NAc, which then causes you to experience pleasure and enjoyment. If you accidentally drink rotten milk, the VTA registers this as a bad thing and tells the NAc, which causes you to experience revulsion, disgust, nausea, practically anything the brain can do to ensure you get the message, 'Do *not* do that again!' This system, when taken together, is the mesolimbic reward pathway.

'Reward' in this context means the positive, pleasurable feelings experienced when we do something our brain approves of. Typically, these are biological functions, like eating food if hungry, or when said foods are nutrient or resource rich (carbohydrates are a valuable energy source as far as the brain is concerned, hence they can be so difficult to resist for dieters). Other things cause much stronger activation of the reward system: things like sex; hence people spend a lot of time and effort to obtain it, despite the fact that we can live without it. Yes, we can.

It doesn't even have to be anything so essential or vivid. Scratching a particularly persistent itch gives pleasurable satisfaction, which is mediated by the reward system. It's the brain telling you that what just happened was good, you should do it again.

In the psychological sense, a reward is a (subjectively) positive response to an occurrence, one that potentially leads to a change in behaviour, so what constitutes a reward can vary considerably. If a rat presses a lever and gets a bit of fruit, it'll press the lever more, so the fruit is a valid reward.[25]

But if instead of fruit it gets the latest Playstation game, it is unlikely to press the lever more frequently. Your average teenager might disagree, but to a rat a Playstation game is of no use or motivational value, so it's not a reward. The point of this is to emphasise that different people (or creatures) find different things rewarding – some people like being scared or unnerved, while others don't and can't see the appeal.

There are several methods via which fear and danger can become 'desirable'. To begin with, we are inherently curious. Even animals such as rats have a tendency to explore something novel when presented with the opportunity. Humans even more so.[26] Consider how often we do something just to see what happens? Anyone who has children will certainly be familiar with this often-destructive tendency. We are drawn to novelty value. We are faced with a huge variety of new sensations and experiences, so why go for the ones that involve fear and danger, two bad things, rather than the many benign-but-equally-unfamiliar ones?

The mesolimbic reward pathway provides pleasure when you do something good. But 'something good' covers a very wide range of possibilities, and this includes *when something bad stops happening*. Due to adrenalin and the fight-or-flight response, periods of fear and terror are incredibly vivid, where all your senses and systems are alert and poised for danger. But, usually, the source of the danger or fear will go away (especially given our overly paranoid brains). The brain recognises that there was a threat, but now it's gone.

You were in a haunted house, and now you're outside. You were hurtling through the air on the way to certain death, but now you're on the ground and alive. You were hearing a terrifying story, but now it's finished and the bloodthirsty serial killer never appeared. In each case, the reward pathway is recognising danger that suddenly ceases, so whatever you did to stop the danger, *it's vitally important that you do that next time*. As such, it triggers a very powerful reward response. In most cases, like eating or sex, you just did something to improve your existence in the short term, but here you *avoided death*! This is far more important. On top of this, with the adrenalin of a fight-or-flight response coursing through our systems everything feels enhanced and heightened. The rush and relief that follows a scare can be intensely stimulating – more so than most other things.

The mesolimbic pathway has important neuronal connections and physical links to the hippocampus and the amygdala, allowing it to emphasise memories for certain occurrences it considers important and attach strong emotional resonance to them.[27] It not only rewards or discourages behaviour when it happens; it makes sure that the memory for the event is also particularly potent.

The heightened awareness, the intense rush, the vivid memories; all of this combined means that the experience of encountering something seriously scary can make someone feel more 'alive' than at any other time. When every other experience seems muted and mundane in comparison, it can be a strong motivator to seek out similar 'highs', just as someone used to drinking double-strength espresso won't find an extra-milky latte especially fulfilling.

And, quite often, it has to be a 'genuine' thrill, rather than a synthetic one. The conscious, thinking parts of our brain might be easily fooled in many cases (many of them covered in this book), but they're not *that* gullible. As such, a video game where you drive a high-speed vehicle, no matter how visually realistic, can't hope to provide the same rush and sensation as actually doing it. The same goes for fighting zombies or piloting starships; the human brain recognises what's real and not real, and can cope with the distinction, despite what the old 'video games lead to violence' arguments suggest.

But if realistic video games aren't scary, how are totally abstract things like stories in books so terrifying? It may be to do with control. When playing a video game, you are in total control of the environment; you can pause the game, it responds to your actions in it, and so on. This isn't the case for scary books or films, where the individual is a passive observer and, while caught up in the narrative, has no influence over what happens in it. (You can close a book, but that doesn't alter the story.) Sometimes the impressions and experiences of the film or book can stay with us long after, unsettling us for quite some time. The vivid memories will explain this, as they keep being revisited and activated as they 'bed in'. Overall, the more the brain retains control over events, the less scary they are. This is why some things that are 'best left to the imagination' are actually more terrifying than the goriest effects.

The 1970s, long before CGI and advanced prosthetics, are widely regarded by connoisseurs of the genre as a golden age of horror films. All the scares had to come from suggestion, timing, atmosphere and other clever tricks. As a result, the brain's tendency to look for and predict threats and dangers did most of the work, causing people literally to jump at shadows. The arrival of cutting-edge effects via big Hollywood studios meant the actual horror was far more blatant and direct, with buckets of blood and CGI replacing psychological suspense. There's room for both approaches, and others, but when the horror is conveyed so directly, the brain isn't as engaged, leaving it free to think and analyse, and remain aware that this is all a fictional scenario that could be avoided at any time, and as such the scares don't have the same impact. Video game makers have learned this, with survival horror games being a genre that requires the character to avoid an overwhelming danger in a tense, uncertain environment, rather than blow it into countless wobbly pieces with an oversized laser cannon.[28]

It's arguably the same with extreme sports and other thrill-seeking activities. The human brain is perfectly able to distinguish actual risk from artificial risk, and there usually needs to be the very real possibility of unpleasant consequences for the true thrill to be experienced. A complex set-up using screens, harnesses and giant fans could feasibly replicate the sensation of bungee jumping, but it would be unlikely to be authentic enough to convince your brain that you are falling from a great height, and thus the danger of actually hitting the ground is removed, and the experience is not the same. The perception of travelling up and down quickly through space is hard to replicate without actually doing it, hence the existence of rollercoasters.

The less control you have over the scary sensation, the more thrilling it is. But there's a cut-off point, as there still has to be some influence over events in order to make it 'fun' scary, rather than simply terrifying. Falling out of a plane with a parachute is considered exciting and fun. Falling out of a plane *without* a parachute on your back is not. For the brain to *enjoy* a thrilling activity, it seems there has to be some actual risk involved, but also some ability to influence the outcome, so the risks can be avoided. Most people who survive a car crash feel relieved to be alive, but there's rarely any desire to go through it again.

Also, the brain has that weird habit, hinted at earlier, called counterfactual thinking; the tendency to dwell on the possible negative outcomes of events *that never happened.*[29] This is going to be even more noticeable when the event itself was a scary one, as there's the sense of actual danger. If you narrowly avoid being hit by a car while crossing the road, you might think about how you *could* have been hit for days afterwards. But you weren't; nothing has physically changed for you at all. But the brain really does like to focus on a potential threat, be it in the past, present or future.

People who enjoy this sort of thing are often labelled adrenalin junkies. 'Sensation seeking' is a recognised personality trait,[30] where individuals constantly strive for new, varied, complex and intense experiences, invariably at some physical/financial/legal risk (losing money and getting arrested are also dangers many people strongly wish to avoid). The previous paragraphs argued that a certain amount of control over events is required to enjoy thrills properly, but it's possible that sensation-seeking tendencies cloud the ability to assess or recognise risk and control accurately. A psychological study from the late 1980s looked at skiers, comparing injured skiers to uninjured skiers.[31] They found injured skiers were far more likely to be sensation seekers than the uninjured ones, suggesting their drive for thrilling sensations caused them to make decisions or perform actions that pushed events beyond their ability to control, resulting in injury. It's a cruel irony that a desire for seeking risk may also cloud your ability to recognise it.

Why some people end up with such extreme tendencies is uncertain. It could just happen gradually, a brief flirtation with a risky experience providing some enjoyable thrills, leading to seeking out more and more with ever increasing intensity. This is the traditional 'slippery slope' argument. Quite an appropriate term for skiers, really.

Some studies have looked into more biological or neurological factors. There's some evidence that certain genes, such as *DRD4*, which encodes a certain class of dopamine receptor, can be mutated in sensation-seeking individuals, suggesting that activity in the mesolimbic reward pathway is altered, resulting in changes in the way sensations are rewarded.[32] If the mesolimbic pathway is more active, intense experiences may be even more powerful. But if it is less powerful, it may require more intense stimulation to achieve true enjoyment as a result; the sort of thing most of us take for granted would require extra life-risking effort. Either way, people could end up seeking more stimulation. Trying to figure out the role of a specific gene in the brain is always a long and complex process, so we don't know this for certain yet.

Another study from 2007 by Sarah B. Martin and her colleagues scanned the brains of dozens of subjects with varying scores on the experience-seeking personality scale and their paper claims that sensation-seeking behaviour is correlated with an enlarged right anterior hippocampus.[33] The evidence suggests that this is the part of the brain and memory system that is responsible for processing and recognising novelty. Basically, the memory system runs information via this area and says, 'Have a look at this. Have we seen this before?' and the right anterior hippocampus says yes or no. We don't know exactly what the increased size of this area means. It could be that the individual has experienced so many novel things that the novelty-recognising area has expanded to cope, or maybe it's that the novelty-detecting region is overly developed so requires something a lot more unusual to be truly recognised as novel. If this were the case, novel stimulations and experiences are potentially more important and salient to these individuals.

Whatever the actual cause for this anterior hippocampal enlargement, for a neuroscientist it's actually quite cool to see something as complex and subtle as a personality trait potentially reflected by visible physical differences in the brain. It doesn't happen nearly as often as the media implies.

Overall, some people actually enjoy the experience of encountering something that causes fear. The fight-or-flight response triggered by this leads to a wealth of heightened experiences occurring in the brain (and the palpable relief that occurs when it ends), and this can be exploited for entertainment purposes within certain parameters. Some people may have subtle differences in brain structure or function that cause them to seek out these intense risk- and fear-related sensation, to sometimes alarming extents. But that's nothing to pass judgement on; once you get past the overall structural consistencies, everyone's brain is different, and those differences are nothing to be afraid of, even if you do enjoy being afraid of things.

You look great - it's nice when people don't worry about their weight

(Why criticism is more powerful than praise)

'Sticks and stones will break my bones, but names will never hurt me.' This claim doesn't really stand up to much scrutiny, does it? Firstly, the hurt caused by a broken bone is obviously quite extreme, so shouldn't be used as a casual baseline for pain. Secondly, if names and insults genuinely don't hurt at all, why does this saying even exist? There's no similar saying to point out that, 'Knives and blades will slash you up but marshmallows are pretty harmless.' Praise is very nice but, let's be honest, criticism *stings*.

Taken at face value, the title of this section is a compliment. If anything, it's actually two compliments, as it flatters both appearance and attitude. But it is unlikely that the person it's directed at will interpret it as such. The criticism is subtle and requires some working out, as it is mostly implied. Despite this, it is the criticism that becomes the stronger element. This is just one of countless examples of a phenomenon that arises from the workings of our brains; criticism typically carries more weight than praise.

If you've ever had a new haircut or outfit or told a funny story to a group or anything else like this, it doesn't matter how many people praise your look or laugh at your jokes, it's the ones who hesitate before saying something nice or roll their eyes wearily at you that will stick with you and make you feel bad.

What's happening here? If it's so unpleasant, why do our brains take criticism so seriously? Is there an actual neurological mechanism for it? Or is it just some morbid psychological fascination with unpleasantness, like the bizarre urge to pick at a scab or poke a loose tooth? There is, of course, more than one possible answer.

To the brain, bad things are typically more potent than good things.[34] At the very fundamental neurological level, the potency of criticism may be due to the action of the hormone cortisol. Cortisol is released by the brain in response to stressful events; it is one of the chemical triggers of the fight-or-flight response, and is widely regarded as the cause of all the issues brought about by constant stress. Its release is controlled mainly by the hypothalamic–pituitary–adrenal (HPA) axis, which is a complex connection of neurological and endocrine (meaning hormone-regulating) areas of the brain and body that coordinate the general response to stress. It was previously believed that the HPA axis was activated in response to a stressful event of any sort, such as a sudden loud noise. But later research found it was a bit more selective than that and was activated only under certain conditions. One theory today is that the HPA axis is activated only when a 'goal' is threatened.[35] For example, if you're walking along and some bird droppings land on you, that's annoying and arguably harmful for hygiene reasons, but it's unlikely to activate the HPA mediated response because 'not being soiled by an errant bird' wasn't really a conscious goal of yours. But if the same bird were to target you while you're walking to a very important job interview, then it is very likely to trigger the HPA response, because you had a definite goal: go to the job interview, impress them, get the job. And now it's been largely thwarted. There are many schools of thought about what to wear to a job interview, but 'a generous layer of avian digestion by-product' doesn't feature in any of them.

The most obvious 'goal' is self-preservation, so if your goal is to stay alive and something occurs that might interfere with your goal by stopping you being alive, the HPA axis would activate the stress response. This is part of the reason it was believed the HPA response responded to anything, because humans can and do see threats to the self everywhere.

However, humans are complex, and one result of this is they rely on the opinions and feedback of other humans to a considerable degree. The social self-preservation theory states that humans have a deep-rooted motivation to preserve their social standing (to continue being liked by the people whose approval they value). This gives rise to social-evaluative threat. Specifically, anything that threatens someone's perceived social standing or image interferes with the goal of being liked, and therefore activates the HPA axis, releasing cortisol in the system.

Criticisms, insults, rejections, mockery, these attack and potentially damage our sense of self-worth, especially if done publicly, which interferes with our goal of being liked and accepted. The stress this causes releases cortisol, which has numerous physiological effects (such as increasing release of glucose), but also has direct effects on our brain. We are aware of how the fight-or-flight response heightens our focus and makes our memories more vivid and prominent. Cortisol, along with other hormones released, potentially causes this to happen (to varying degrees) when we're criticised; it makes us experience an actual physical reaction that sensitises us and emphasises the memory of the event. This whole chapter is based on the brain's tendency to go overboard when looking for threats, and there's no real reason why this wouldn't include criticism. And when something negative happens and we experience it first hand, producing all the relevant emotions and sensations, the hippocampus and amygdala processes spark into life again, and end up emotionally enhancing the memory and storing it more prominently.

Nice things, such as receiving praise, also produce a neurological reaction via the release of oxytocin, which makes us experience pleasure, but in a less potent and more fleeting manner. The chemistry of oxytocin means it's removed from the bloodstream in about five minutes; cortisol, by contrast, can linger for over an hour, maybe even two, so its effects are far more persistent.[36] The fleeting nature of pleasure signals may seem a bit of a harsh move by nature, but when things cause us intense pleasure for long periods they tend to be quite incapacitating, as we'll see later.

However, it's easy but misleading to attribute everything that goes on in the brain to the actions of specific chemicals, and this is something that more 'mainstream' neuroscience reports do often. Let's look at some other possible explanations for this emphasis of criticism.

Novelty may also play a role. Despite what online comment sections might suggest, most people (with some cultural variations, admittedly) interact with others in a respectful manner due to social norms and etiquette; shouting abuse at someone in the street is not something that respectable people do, unless it's directed at traffic wardens, who are apparently exempt from this rule. Consideration and low-level praise are the norm, like saying thank you to the cashier for handing you your change even though it's your money and they've no right to keep it. When something becomes the norm, our novelty-preferring brains start to filter it out more often via the process of habituation.[37] Something happens all the time, so why waste precious mental resources focusing on it when it's safe to ignore?

Mild praise is the standard, so criticism is going to have more impact purely because it's atypical. The single disproving face in a laughing audience is going to stand out more *because* it's so different. Our visual and attention systems have developed to focus on novelty, difference and 'threat', all of which are technically embodied by the grumpy-looking person. Similarly, if we're used to hearing 'well done' and 'good job' as meaningless platitudes, then someone saying, 'You were crap!' is going to be all the more jarring because it doesn't happen as often. And we shall dwell on an unpleasant experience all the more to figure out why it happened, so we can avoid it next time.

Chapter 2 discussed the fact that the workings of the brain tend to make us all somewhat egotistical, with a tendency to interpret events and remember things in such a way as to give us a better self-image. If this is our default state, praise is just telling us what we already 'know', whereas direct criticism is harder to misinterpret and a shock to the system.

If you put yourself 'out there' in some form, via a performance, created material or just an opinion you think is worthy of sharing, you are essentially saying, 'I think you will like this'; you're visibly seeking people's approval. Unless you're alarmingly confident then there's always an element of doubt and awareness of the possibility that you are wrong. In this instance you are sensitive to the risk of rejection, primed to look for any signs of disapproval or criticism, especially if it's regarding something that you take great pride in or that required a lot of time and effort. When you're primed to look for something you're worried about, you're more likely to find it. Just as a hypochondriac is always able to find himself showing worrying symptoms for rare diseases. This process is called confirmation bias – we seize on what we're looking for and ignore anything that doesn't match up to it.[38]

Our brains can really make judgements based only on what we know, and what we know is based on our own conclusions and experiences, so we tend to judge people's actions based on what we do. So if we're polite and complimentary just because social norms say we should be, then surely everyone else does the same? As a result, every item of praise you receive can be somewhat dubious as to whether it's genuine or not. But if someone criticises you, not only were you bad, you were *so* bad that someone was willing to go against social norms to point it out. And thus, once again, criticism carries more weight than praise.

The brain's elaborate system for identifying and responding to potential threats may well have enabled humankind to survive the long periods in the wilderness and become the sophisticated, civilised species we are today, but it's not without drawbacks. Our complex intellects allow us not only to identify threats but to anticipate and imagine them too. There are many ways to threaten or frighten a human, which leads to the brain responding neurologically, psychologically or sociologically.

This process can, depressingly, cause vulnerabilities that other humans are able to take advantage of, resulting in actual threats, in a sense. You may be familiar with 'negging', a tactic used by pick-up artists where they approach women and say something that sounds like a compliment but is actually meant to criticise and insult. If a man approached a woman and said the title of this section, that would be negging. Or he might say something like, 'I like your hair – most women with your face wouldn't risk a style like that', or, 'I normally don't like girls as short as you, but you seem cool', or, 'That outfit will look great once you lose some weight', or, 'I've no clue how to speak to women because I've only ever seen them through binoculars so I'm going to use cheap psychological trickery on you in the hope that I will do enough damage to your self-confidence that you are willing to sleep with me.' That last one isn't a typical negging line, admittedly, but in truth it's what they're all saying.

It doesn't need to be this sinister, though. We probably all know the type of person who, when someone has done something to be proud of, will immediately jump in to point out the bits they did wrong. Because why go to the effort of achieving something yourself when you can just bring others down to make yourself feel better?

It's a cruel irony that in looking for threats so diligently, the brain ends up actually creating them.

4

Think you're clever, do you?

The baffling science of intelligence

What makes the human brain special or unique? There are numerous possible answers, but the most likely is that it provides us with superior intelligence. Many creatures are capable of all the basic functions our brain is responsible for, but thus far no other known creature has created its own philosophy, or vehicles, or clothing, or energy sources, or religion, or a *single* type of pasta, let alone over three hundred varieties. Despite the fact that this book is largely about the things the human brain does inefficiently or bizarrely, it's important not to overlook the fact that it's clearly doing something right if it's enabled humans to have such a rich, multifaceted and varied internal existence, and achieve as much as they have.

There's a famous quote that says, 'If the human brain were so simple that we could understand it, we would be so simple that we couldn't.' If you look into the science of the brain and how it relates to intelligence, there's a strong element of truth in this aphorism. Our brains make us intelligent enough to recognise that we *are* intelligent, observant enough to realise this isn't typical in the world, and curious enough to wonder why this is the case. But we don't yet seem to be intelligent enough to grasp easily where our intelligence comes from and how it works. So we have to fall back on studies of the brain and psychology to get any idea of how the whole process comes about. Science itself exists thanks to our intelligence, and now we use science to figure out how our intelligence works? This is either very efficient or circular reasoning, I'm not smart enough to tell.

Confusing, messy, often contradictory, and hard to get your head round; this is as good a description of intelligence itself as any you're likely to find. It's difficult to measure and even define reliably but I'm going to go through how we use intelligence and its strange properties in this chapter.

My IQ is 270 . . . or some other big number
(Why measuring intelligence is harder than you think)

Are you intelligent?

Asking yourself that means the answer is definitely yes. It demonstrates you are capable of many cognitive processes that automatically qualify you for the title of 'most intelligent species on earth'. You are able to grasp and retain a concept such as intelligence, something that has no set definition and no physical presence in the real world. You are aware of yourself as an individual entity, something with a limited existence in the world. You are able to consider your own properties and abilities and measure them against some ideal but currently-not-existing goal or deduce that they may be limited in comparison to those of others. No other creature on earth is capable of this level of mental complexity. Not bad for what is basically a low-level neurosis.

So humans are, by some margin, the most intelligent

species on earth. What does that *mean*, though? Intelligence, like irony or daylight-saving time, is something most people have a basic grasp of but struggle to explain in detail.

This obviously presents a problem for science. There are many different definitions of intelligence, provided by many scientists over the decades. French scientists Binet and Simon, inventors of one of the first rigorous IQ tests, defined intelligence as: 'To judge well, to comprehend well, to reason well; these are the essential activities of intelligence.' David Weschler, an American psychologist who devised numerous theories and measurements of intelligence, which are still used today via tests such as the Weschler Adult Intelligence Scale, described intelligence as 'the aggregate of the global capacity to act purposefully, to deal effectively with the environment'. Philip E. Vernon, another leading name in the field, referred to intelligence as 'the effective all-round cognitive abilities to comprehend, to grasp relations and reason'.

But don't go thinking it's all pointless speculation; there are many aspects of intelligence that are generally agreed on: it reflects the brain's ability to do . . . stuff. More precisely, the brain's ability to handle and exploit information. Terms such as reasoning, abstract thought, deducing patterns, comprehension; things like this are regularly cited as examples of superior intelligence. This makes a certain logical sense. All of these typically involve assessing and manipulating information on an entirely intangible basis. Simply put, humans are intelligent enough to work things out without having to interact with them directly.

For example, if a typical human approaches a gate held shut with large padlocks, they'll quickly think, 'Well, that's locked', and go find another entrance. This may seem trivial, but it's a clear sign of intelligence; the person observes a situation, deduces what it means, and responds accordingly. There is no physical attempt to open the gate, at which point they'd discover, 'Yep, that's locked'; they don't *have* to. Logic, reasoning, comprehension, planning; these have all been utilised to dictate actions. This is intelligence. But that doesn't clarify how we study and measure intelligence. Manipulating information in complex ways inside the brain is all well and good, but it's not something that can be observed directly (even the most advanced brain scanners just show us blurs of differing colour at present, which isn't especially useful) so measuring it can be done only indirectly by observing behaviour and performance on specially designed tests.

At this point, you might think that something major has been missed here, because we *do* have a way of measuring intelligence: IQ tests. Everyone knows about IQ, meaning Intelligence Quotient; it's a measurement of how smart you are. Your mass is provided by measuring your weight; your height is determined by measuring how tall you are; your intoxication level is calculated by breathing into one of those gadgets the police make you breathe into; and your intelligence is measured by IQ tests. Simple, right?

Not exactly. IQ is a measurement that takes the slippery, unspecified nature of intelligence into account, but most people assume it's far more definitive than it is. Here's the important fact you need to remember: the average IQ of a population is 100. *Without exception*. If someone says, 'The average IQ of [country x] is only 85', then this is wrong. It's basically the same as saying, 'The length of a metre in [country x] is only 85 cm'; this is logically impossible, and the same is true for IQ.

Legitimate IQ tests tell you where you fall within the typical distribution of intellect in your population, according to a proposed 'normal' distribution. This normal distribution dictates that the 'mean' IQ is 100. An IQ between 90 and 110 is classed as average, between 110 and 119 is 'high average', between 120 and 129 is 'superior', and anything over 130 is 'very superior'. Conversely, an IQ between 80 and 89 is 'low average', 70 to 79 is 'borderline', and anything below 69 is considered 'extremely low'.

Using this system, over 80 per cent of the population will fall in the average zones, with IQs ranging from 80 to 110. The further out you go, the fewer people you'll find with these IQs; less than 5 per cent of the population will be have a very superior or extremely low IQ. A typical IQ test doesn't directly measure your raw intelligence, but reveals how intelligent you are compared to the rest of the population.

This can have some confusing consequences. Say a potent but bizarrely specific virus wiped out everyone in the world with an IQ of over 100. The people left behind would *still have an average IQ of 100*. Those with IQs of 99 before the plague hit would now suddenly have IQs of 130+ and be classed as the *crème de la crème* of the intellectual elite. Think of it in terms of currency. In Britain the value of the pound fluctuates in accordance with what happens in the economy, but there are always 100 pennies to the pound, so the pound has values that are both flexible and fixed. IQ is basically the same: the average IQ is always 100, but what an IQ of 100 is actually worth in terms of intelligence is variable.

This normalisation and adhering to population averages means that IQ measurement can be a bit restrictive. People such as Albert Einstein and Stephen Hawking reportedly have IQs in the region of 160, which is still very superior but doesn't sound so impressive when you consider the population average is 100. So if you meet someone who does claim to have an IQ of 270 or some such, they're probably wrong. They've been using some alternative type of test that isn't considered scientifically valid, or they've seriously misread their results, which does undermine their claim to be a super genius.

This isn't to say that such IQs don't exist at all; some of the most intelligent people on record supposedly had IQs of over 250, according to the *Guinness Book of Records*, although the category of Highest IQ was retired in 1990 due to the uncertainty and ambiguity of the tests at this level.

The IQ tests used by scientists and researchers are meticulously designed; they're used as actual tools, like microscopes and mass spectrometers. They cost a lot of money (so aren't given away online for free). The tests are designed to assess normal, average intelligences in the widest possible range of people. As a result, the further to the extremes you go, the less useful they tend to be. You can demonstrate many concepts of physics in the school classroom with everyday items (for instance, using weights of different sizes to show the constant force of gravity, or a spring to show elasticity) but, if you delve into complex physics, you need particle accelerators or nuclear reactors and frighteningly complex mathematics.

So it is when you have someone of extremely high intelligence; it just becomes much harder to measure. These scientific IQ tests measure things such as spatial awareness with pattern completion tests, comprehension speeds with dedicated questions, verbal fluency by getting the subject to list words from certain categories, and stuff like that; all reasonable things to look into but not something that is likely to tax a super genius to the extent where it would be possible to spot the very limits of his or her intelligence. It's a bit like using bathroom scales to weigh elephants; they can be useful for a standard range of weights, but at this level they'll give no useful data, just a load of broken plastic and springs.

Another concern is that intelligence tests claim to measure intelligence, and we know what intelligence is because intelligence tests tell us. You can see why some of the more cynical scientist types wouldn't be happy with this situation. In truth, the more common tests have been revised repeatedly and assessed for reliability often, but some still feel that this is just ignoring the underlying problem.

Many like to point out that performance on intelligence tests is actually more indicative of social upbringing, general health, aptitude to testing, education level, and so on. Things that aren't intelligence, in other words. So the tests may be useful, but not for what they're intended.

It's not all doom and gloom. Scientists aren't ignorant of these criticisms and are a resourceful bunch. Today, intelligence tests are more useful – they provide a wide range of assessments (spatial awareness, arithmetic etc.), rather than one general assessment, and this gives us a more robust and thorough demonstration of ability. Studies have shown that performance on intelligence tests also seems to remain fairly stable over a person's lifetime despite all the changes or learning they experience, so they must be detecting some inherent quality rather than just random circumstance.[1]

So, now you know what we know, or what we think we know. One of the generally accepted signs of intelligence is an awareness and acceptance of what you don't know. Good job.

Where are your trousers, professor? (How intelligent people end up doing stupid things)

The stereotype of an academic is a white-haired white-coated chap (it's almost always a man) in late middle age, talking quickly and often about his field of study while being utterly clueless about the world around him, effortlessly describing the fruit fly genome while absent-mindedly buttering his tie. Social norms and day-to-day tasks are completely alien and baffling to him; he knows everything there is to know about his subject, but little to nothing beyond that.

Being intelligent isn't like being strong; a strong person is strong in every context. However, someone brilliant in one context can seem like a shuddering dunce in another.

This is because intelligence, unlike physical strength, is a product of the never uncomplicated brain. So what are the brain processes that underpin intelligence, and why is it so variable? Firstly, there is ongoing debate in psychology about whether or not humans use a single intelligence, or several different types. Current data suggests it is probably a combination of things.

A dominant view is that there is a single property that underpins our intelligence, which can be expressed in varying ways. This is often known as 'Spearman's g', or just g. Named after Charles Spearmen, a scientist who did a great service for intelligence research and science in general in the 1920s by developing factor analysis. The previous section revealed how IQ tests are commonly used despite certain reservations; factor analysis is something that makes them (and other tests) useful.

Factor analysis is a mathematically dense process but what you need to know is that it is a form of statistical decomposition. This is where you take large volumes of data (for example, those produced by IQ tests) and mathematically break them down in various ways and look for factors connecting or influencing the results. These factors aren't known beforehand, but factor analysis can flush them out. If students at a school got middling marks overall in their exams, the headmaster might want to see exactly how the marks were achieved in more detail. Factor analysis could be used to assess the information from all the exam scores and take a closer look. It could reveal that maths questions were generally answered well, but history questions were answered poorly. The headmaster can then feel justified about yelling at the history teachers for wasting time and money (although he probably isn't justified, given the many possible explanations for poor results).

Spearman used a process similar to this to assess IQ tests and discovered that there was seemingly one underlying factor that underpinned test performance. This was labelled the single general factor, *g*, and if there's anything in science that represents what your everyday person would think of as intelligence, it's *g*.

It would be wrong to say that *g* = all possible intelligence, as intelligence can manifest in so many ways. It's more a general 'core' of intellectual ability. It's viewed as something like the foundations and frame of a house. You can add extensions and furniture, but if the underlying structure isn't strong enough it'll be futile. Similarly, you can learn all the big words and memory tricks you like, but if your *g* isn't up to scratch you won't be able to do much with them.

Research suggests there might be a part of the brain that is responsible for *g*. Chapter 2 discussed short-term memory in detail and alluded to the term 'working memory'. This refers to the actual processing and manipulation, the 'using' of the information in short-term memory. In the early 2000s, Professor Klaus Oberauer and his colleagues ran a series of tests and found that a subject's performance on working-memory tests corresponded strongly with tests to determine his or her *g*, indicating that a person's working-memory capacity is a major factor in overall intelligence.[2] Ultimately, if you score highly on a working-memory task, you're very likely to score highly on a range of IQ tests. It makes logical sense; intelligence involves obtaining, retaining and using information as efficiently as possible, and IQ tests are designed to measure this. But such processes are basically what the working memory is for.

Scanning studies and investigations of people with brain injuries provide compelling evidence for a pivotal role of the prefrontal cortex in processing both *g* and working memory, with those afflicted with frontal-lobe injury demonstrating a wide range of unusual memory problems, typically traced back to a deficit in working memory, thus further implying a large overlap between the two things. This prefrontal cortex is the right behind the forehead, the beginning of the frontal lobe that is regularly implicated in higher 'executive' functions such as thinking, attention and consciousness.

But working memory and *g* are not the whole story. Working-memory processes mostly work with verbal information, supported by words and terms we could speak aloud, like an internal monologue. Intelligence, on the other hand, is applicable to all types of information (visual, spatial, numerical . . .), prompting researchers to look beyond *g* when trying to define and explain intelligence.

Raymond Cattell (a former student of Charles Spearman) and his student John Horn devised newer methods of factor analysis and identified two types of intelligence in studies spanning the 1940s to 1960s; fluid intelligence and crystallised intelligence.

Fluid intelligence is the ability to *use* information, work with it, apply it, and so on. Solving a Rubik's cube requires fluid intelligence, as does working out why your partner isn't talking to you when you have no memory of doing anything wrong. In each case, the information you have is new and you have to work out what to do with it in order to arrive at an outcome that benefits you.

Crystallised intelligence is the information you have stored in memory and can utilise to help you get the better of situations. Knowing the lead actor in an obscure 1950s film for a pub quiz requires crystallised intelligence. Knowing all the capital cities of the northern hemisphere is crystallised intelligence. Learning a second (or third or fourth) language utilises crystallised intelligence. Crystallised intelligence is the knowledge you have accumulated, where fluid intelligence is how well you can use it or deal with unfamiliar things that need working out.

It's fair to say that fluid intelligence is another variation of *g* and working memory; the manipulation and processing of information. But crystallised intelligence is increasingly viewed as a separate system, and the workings of the brain back this up. One quite telling fact is that fluid intelligence declines as we age; someone aged eighty will perform worse on a fluid intelligence test than he or she did aged thirty, or fifty. Neuroanatomical studies (and numerous autopsies) revealed the prefrontal cortex, believed responsible for fluid intelligence, atrophies more with age than most other brain regions.

Contrastingly, crystallised intelligence remains stable over a lifetime. Someone who learns French at eighteen will still be able to speak it at eighty-five, unless they stopped using it and forgot it at nineteen. Crystallised intelligence is supported by long-term memories, which are distributed widely throughout the brain and tend to be resilient enough to withstand the ravages of time. The prefrontal cortex is a demanding energetic region that needs to engage in constant active processing to support fluid intelligence, actions that are quite dynamic and thus more likely to result in gradual wear and tear (intense neuronal activity tends to give off a lot of waste products such as free radicals, energetic particles that are harmful to cells).

Both types of intelligence are interdependent; there's no point in being able to manipulate information if you can't access any of it, and vice versa. It's tricky to separate them clearly for study. Luckily, intelligence tests can be designed to focus mostly on either fluid or crystallised intelligence. Tests that require individuals to analyse unfamiliar patterns and identify odd ones out or work out how they are interconnected are thought to assess fluid intelligence; all the information is novel and needs to be processed, so crystallised-intelligence use is minimal. Similarly, tests of recall and knowledge such as word-list memory, or the aforementioned pub quizzes, focus on crystallised intelligence.

It's never quite *that* simple of course. Tasks where you have to sort unfamiliar patterns still rely on an awareness of images, colours, even the means by which you complete the test (if it's rearranging a series of cards, you'll be using your knowledge of what cards are and how to arrange them). This is another thing that makes brain-scanning studies tricky; even doing a simple task involves multiple brain regions. But, in general, tasks for fluid intelligence tend to show greater activity in the prefrontal cortex and associated regions, and crystallised intelligence tasks suggest a role of the wider cortex, often the parietal-lobe (the upper-middle bit of the brain) regions, such as the supramarginal gyrus and Broca's area. The former is often thought of as being necessary for storage and processing of information concerning emotion and some sensory data, while the latter is a key part of our language-processing system. Both are interconnected, and suggest functions requiring access to long-term memory data. While it's still not clear cut, there's mounting evidence to support this fluid/crystallised distinction of general intelligence.

Miles Kingston captures the theory brilliantly: 'Knowledge is knowing that a tomato is a fruit; wisdom is not putting it in a fruit salad.' It requires crystallised intelligence to know how a tomato is classed, and fluid intelligence to apply this information when making a fruit salad. You might now think that fluid intelligence sounds a lot like common sense. Yes, that would be another example. But, for some scientists, two distinct types of intelligence are still not enough. They want more.

The logic is that a single general intelligence is insufficient for explaining the wide variety of intellectual abilities humans can demonstrate. Consider footballers – they often didn't thrive academically, but being able to play a complicated sport like football at professional level requires a great deal of intellectual ability such as precise control, calculating force and angles, spatial awareness of a wide area, and so on. Concentrating on your job while filtering out the rantings of the obsessive fans takes considerable mental fortitude. The common concept of 'intelligence' is clearly a bit restrictive.

Perhaps the starkest examples are 'savants', individuals with some form of neurological disorder, who show an extreme affinity or ability for complex tasks involving maths, music, memory, etc. In the film *Rain Man*, Dustin Hoffman plays Raymond Babbit, an autistic but mathematically gifted psychiatric patient. The character was inspired by a real individual called Kim Peek who was dubbed a 'mega-savant' for his ability to memorise, to the word, up to twelve thousand books.

These examples and more lead to the development of multiple intelligence theories, because how can someone be both unintelligent in one sphere and a gifted in another if there's only one type of intelligence? The earliest theory of this nature is probably that put forward by Louis Leon Thurstone in 1938, who proposed that human intelligence was made up of seven Primary Mental Abilities:

Verbal comprehension (understanding words: 'Hey, I know what that means!')
Verbal fluency (using language: 'Come here and say that, you acephalous buffoon!')
Memory ('Wait, I remember you, you're the cage-fighting world champion!')
Arithmetic ability ('The odds of me winning this fight are about 82523 to 1.')
Perceptual speed (spotting and linking details: 'Is he

wearing a necklace made of human teeth?')
Inductive reasoning (deriving ideas and rules from situations: 'Any attempt to placate this beast is only going to anger him further.')
Spatial visualisation (mentally visualising/manipulating a 3D environment: 'If I tip this table it'll slow him down and I can dive out that window.')

Thurstone derived his Primary Mental Abilities after devising his own methods of factor analysis and applying them to IQ test results of thousands of college students.[3] However, reanalysis of his results using more traditional factor analysis showed there was a single ability influencing all the tests, rather than several different ones. Basically, he'd discovered g again. This and other criticisms (for instance that he studied only college students, hardly the most representative group when it comes to general human intelligence) meant the Primary Mental Abilities weren't that widely accepted.

Multiple intelligences returned in the 1980s via Howard Gardner, a prominent researcher who proposed that there were several modalities (types) of intelligence, and his aptly titled *Theory of Multiple Intelligences*, following research into brain-damaged patients who still retained certain types of intellectual abilities.[4] His proposed intelligences were similar to Thurstone's in some ways, but also included musical intelligence, and personal intelligences (ability to interact well with people, and ability to judge your own internal state).

The multiple-intelligence theory has its adherents though. Multiple intelligences are popular largely because it means everyone can potentially be intelligent, just not in the 'normal' brainy boffin way. This generalisability is also something it's criticised for. If everyone is intelligent, the concept itself becomes meaningless in the scientific sense. It's like giving everyone a medal for showing up at a school sports day; it's nice that everyone gets to feel good, but it does defeat the point of 'sport'.

So far, the evidence for the multiple-intelligence theory remains debatable. The data available is widely regarded as being yet more evidence for *g* or something like it, combined with personal differences and preferences. What this means is that two people who excel, one at music and one at maths, aren't actually demonstrating two different types of intelligence, but the same general intelligence applied to different types of tasks. Similarly, professional swimmers and tennis players use the same muscle groups to practise their sports; the human body doesn't have dedicated tennis muscles. Nonetheless, a champion swimmer can't automatically play top-level tennis. Intelligence is believed to work in similar ways.

Many argue that it is perfectly plausible to have a high *g* but prefer to utilise and apply it in specific ways, which would manifest as different 'types' of intelligence if you look at it in a certain way. Others argue that these supposed different types of intelligence are more suggestive of personal inclinations based on background, tendencies, influences, and so on.

Current neurological evidence still favours the existence of *g* and the fluid/crystallised set-up. Intelligence in the brain is believed to be due to the way the brain is arranged to organise and coordinate the various types of information, rather than a separate system for each one. This will be covered in more detail later in this chapter.

Current neurological evidence favours the existence of *g* and the fluid/crystallised system. Intelligence is believed to be due to the way the brain is set up to organise and coordinate the various types of information, rather than having a separate system for each one. We all direct our intelligence in certain ways and directions, whether due to preference, upbringing, environment or some underlying bias imparted by subtle neurological properties. This is why you get supposedly very smart people doing things we'd consider daft; it's not that they aren't clever enough to know better, it's that they're too focused elsewhere to care. On the plus side, this probably means it's OK to laugh at them, as they'll be too distracted to notice.

Empty vessels make the most noise
(Why intelligent people can often lose arguments)

One of the most infuriating experiences possible is arguing with someone who's convinced they're right when you know full well that they're wrong, and can prove they're wrong with facts and logic, but still they won't budge. I once witnessed a blazing row between two people, one of whom was adamant that this is the twentieth century, not the twenty-first, because, 'It's *twenty* fifteen? Duh!' That was their actual argument.

Contrast this with the psychological phenomenon known as 'impostor syndrome'. High achievers in many fields persistently underestimate their abilities and achievements despite having *actual evidence* of these things. There are many social elements to this. For instance, it's particularly common in women who achieve success in a traditionally male-dominated environment (aka most of them) so they are likely to be influenced by stereotyping, prejudice, cultural norms and so on. But it's not limited to women, and one of the more interesting aspects is that it predominately affects high achievers – those people with a typically high level of intelligence.

Guess which scientist said this shortly before his death: 'The exaggerated esteem in which my lifework is held makes me very ill at ease. I feel compelled to think of myself as an involuntary swindler.'

Albert Einstein. Not exactly an underachiever.

These two traits, impostor syndrome in intelligent people and illogical self-confidence in less intelligent people, regularly overlap in unhelpful ways. Modern public debate is disastrously skewed due to this. Important issues such as vaccination or climate change are invariably dominated by the impassioned rantings of those who have uninformed personal opinions rather than the calmer explanations of the trained experts, and it's all thanks to a few quirks of the brain's workings.

Basically, people rely on other people as a source of information and support for their own views/beliefs/sense of self-worth, and Chapter 7 on social psychology will go into this in more detail. But, for now, it seems the more confident a person is, the more convincing they are and the more others tend to believe the claims they make. This has been demonstrated in a number of studies, including those conducted in the 1990s by Penrod and Custer, who focused on courtroom settings. These studies looked at the degree to which jurors were convinced by witness testimonies and found that jurors were far more likely to favour witnesses who came across as confident and assured than those who seemed nervous and hesitant or unsure of the details of their claim. This was obviously a worrying finding; the content of a testimony being less influential in determining a verdict than the manner in which it is delivered could have serious ramifications for the justice system. And there's nothing to say it's limited to a courtroom setting; who's to say politics isn't similarly influenced?

Modern politicians are media-trained so they can speak confidently and smoothly on any subject for prolonged periods without saying anything of value. Or worse, something downright stupid like, 'They misunderestimated me' (George W. Bush), or, 'Most of our imports come from overseas' (George W. Bush again). You'd assume that the smartest people would end up running things; the smarter a person is, the better job they'd be able to do. But as counterintuitive as it may seem, the smarter a person is, the greater the odds of them being less confident in their views, and the less confident they come across as being, the less they're trusted. Democracy, everyone.

Intelligent sorts may be less confident because there can often be a general hostility to those of the intellectual persuasion. I'm a neuroscientist by training, but I don't tell people this unless directly asked, because I once got the response, 'Oh, think you're *clever*, do you?'

Do other people get this? If you tell someone you're an Olympic sprinter, does anyone ever say, 'Oh, think you're *fast*, do you?' This seems unlikely. But, regardless, I still end up saying things like, 'I'm a neuroscientist, but it's not as impressive as it sounds.' There are countless social and cultural reasons for anti-intellectualism, but one possibility is that it's a manifestation of the brain's egocentric or 'self-serving' bias and tendency to fear things. People care about their social standing and well-being, and someone seeming more intelligent than them can be perceived as a threat. People who are physically bigger and stronger can certainly be intimidating, but it's a known property. A physically fit person is easy to understand; they just go to the gym more, or have been doing their chosen sport for far longer, right? That's how muscles and such work. Anyone could end up like them if they do what they did, if they had the time or inclination.

But someone who is more intelligent than you presents an unknowable quantity, and as such they could behave in ways that you can't predict or understand. This means the brain cannot work out whether they present a danger or not, and in this situation the old 'better safe than sorry' instinct is activated, triggering suspicion and hostility. It's true that a person could also learn and study to become more intelligent as well, but this is far more complex and uncertain than physical improvement. Lifting weights gives you strong arms, but the connection between learning and intelligence is far more diffuse.

The phenomenon of less-intelligent people being more confident has an actual scientific name: the Dunning–Kruger effect. It is named for David Dunning and Justin Kruger of Cornell University, the researchers who first looked into the phenomenon, inspired by reports of a criminal who held up banks after covering his face with lemon juice, because lemon juice can be used as invisible ink, so he thought his face wouldn't show up on camera.[5]

Just let that sink in for a moment.

Dunning and Kruger got subjects to complete a number of tests, but also asked them to estimate how well they thought they had done on the tests. This produced a remarkable pattern: those who performed badly on the tests almost always assumed they'd done much *much* better, whereas those who did well invariably assumed they'd done worse. Dunning and Kruger argued that those with poor intelligence not only lack the intellectual abilities, they also lack *the ability to recognise that they are bad at something*. The brain's egocentric tendencies kick in again, suppressing things that might lead to a negative opinion of oneself. But also, recognising your own limitations and the superior abilities of others is something that itself requires intelligence. Hence you get people passionately arguing with others about subjects they have no direct experience of, even if the other person has studied the subject all their life. Our brain has only our own experiences to go from, and our baseline assumptions are that everyone is like us. So if we're an idiot . . .

The argument is that an unintelligent person actually cannot 'perceive' what it is to be considerably more intelligent. It's basically like asking a colour-blind person to describe a red and green pattern.

It may be that an 'intelligent' has a similar take on the world, but expressed in different ways. If an intelligent person thinks something was easy then they may assume everyone else finds it easy too. They assume their level of competence is the norm, so they assume their intelligence is the norm (and intelligent people tend to find themselves in jobs and social situations where they're surrounded by other similar types, so they are likely to have a lot of evidence to support this).

But if intelligent people are generally used to learning new things and acquiring new information, they're more likely to be aware that they *don't* know everything and how much there is to know about any given subject, which would undercut confidence when making claims and statements.

For example, in science, you (ideally) have to be painstakingly thorough with your data and research before making any claims as to how something works. A consequence of surrounding yourself with similarly intelligent people means if you do make a mistake or a grandiose claim, they're more likely to spot it and call you on it. A logical consequence of this would be a keen awareness of the things you don't know or aren't sure about, which is often a handicap in a debate or an argument.

These occurrences are common enough to be familiar and problematic, but obviously aren't absolute; not every intelligent person is racked with doubt, and not every less-intelligent person is a self-aggrandising buffoon. There are plenty of intellectuals who are so in love with the sound of their own voice that they genuinely charge people thousands to hear it, and there are ample less-intelligent people who freely admit their limited mental powers with grace and humility. It may also have a cultural aspect; the studies behind the Dunning–Kruger effect almost always focus on Western societies, but some East Asian cultures have shown very different patterns of behaviour, and one explanation put forward for this is these cultures adopt the (healthier) attitude that a lack of awareness is an opportunity for improvement, so the priorities and behaviours are very different.[6]

Are there actual brain regions behind this kind of phenomenon? Is there a part of the brain responsible for working out: 'Am I any good at this thing that I'm doing?' Amazing as it may seem, there might well be. In 2009, Howard Rosen and his colleagues tested a group of about forty patients with neurodegenerative diseases and concluded that accuracy in self-appraisal correlated with the volume of tissue in the right ventromedial (lower part, towards the middle) region of the prefrontal cortex.[7] The study argues that this area of the prefrontal cortex is needed for the emotional and physiological processing required when evaluating your own tendencies and abilities. This is relatively consistent with the accepted functioning of the prefrontal cortex, which is largely all to do with processing and manipulating complex information and working out the best possible opinion of it and response to it .

It's important to note that this study in and of itself is not conclusive; forty patients isn't really enough to say that the data obtained from them is relevant to everyone ever. But research into this ability to assess your own intellectual performance accurately, known as a 'metacognitive ability' (thinking about thinking, if that makes sense), is considered to be quite important, as an inability to perform accurate self-appraisal is a well-known feature of dementia. This is particularly true of frontotemporal dementia, a variation of the disorder that attacks largely the frontal lobe, where the prefrontal cortex is. Patients with this condition often show an inability to assess their performance on a wide variety of tests accurately, which would suggest their ability to assess and evaluate their performance has been seriously compromised. This wide-ranging inability to judge one's performance accurately isn't seen in other types of dementia that damage different brain regions, suggesting an area of the frontal lobe is heavily involved in self-appraisal. So this adds up.

Some propose that this is one reason why dementia patients can turn quite aggressive; they are unable to do things but cannot understand or recognise why, which must be nothing short of enraging.

But even without a neurodegenerative disorder and while

in possession of a fully functioning prefrontal cortex, this means only that you are capable of self-appraisal; there's nothing to say your self-appraisal will be correct. Hence we end up with confident clowns and insecure intellectuals. And it's apparently human nature that we pay more attention to the confident ones.

Crosswords don't actually keep your brain sharp

(Why it's very difficult to boost your brain power)

There are many ways to *appear* more intelligent (using pompous terms such as '*au courant*', carrying *The Economist*), but can you *actually become* more intelligent? Is it possible to 'boost your brain power'?

In the sense of the body, power usually means the ability to do something or act in a particular way, and 'brain power' is invariably linked to abilities that would come under the heading of intelligence. You could feasibly increase the amount of *energy* contained within your brain by using your head to complete a circuit connected to an industrial generator, but that's not going to be something that benefits you, unless you're especially keen to have your mind literally blown (to bits).

You've probably seen ads for things that claim to offer substances, tools or techniques for boosting your brain power, usually for a price. It's highly unlikely that any of these things will actually work in any significant way, because if they did they'd be far more popular, with everyone getting smarter and bigger-brained until we're all crushed under the weight of our own skulls. But how does one genuinely increase brain power, boosting intelligence?

For this, it would be useful to know what differentiates the unintelligent brain from the intelligent one, and how do we turn the former into the latter? One potential factor is something that seems completely wrong: intelligent brains apparently use *less* power.

This counterintuitive argument is something that arose from scanning studies directly observing and recording brain activity, such as functional magnetic resonance imaging (fMRI). This is a clever technique where people are placed in MRI scanners and their metabolic activity (where the tissues and cells in the body are 'doing stuff') is observed. Metabolic activity requires oxygen, supplied by the blood. An fMRI scanner can tell the difference between oxygenated blood and deoxygenated blood and when one becomes the other, which occurs at high levels in areas of the body that are metabolically active, like brain regions working hard at a task. Basically, fMRI can monitor brain activity and spot when one part of the brain is especially active. For example, if a subject is doing a memory task, the areas of the brain required for memory processing will be more active than usual, and this shows up on the scanner. Areas showing increased activity would be identifiable as memory-processing areas.

It isn't as simple as that because the brain is constantly active in many different ways, so finding the 'more' active bits requires much filtering and analysis. However, the bulk of modern research about identifying brain regions that have specific functions have utilised fMRI.

So far, so good; you'd expect that a region responsible for a specific action would be more active when having to do that action, like a weightlifter's bicep is using more energy when picking up a dumb-bell. But no. Bizarre findings from several studies, such as those from Larson and others in 1995,[8] showed that in tasks designed to test fluid intelligence, activity was seen in the prefrontal cortex . . . except when the subject was *very good* at the task.

To clarify, the region supposedly responsible for fluid intelligence apparently wasn't used in people who had high levels of fluid intelligence. This didn't make a lot of sense – like weighing people and finding that only lighter people show up on the scales. Further analysis found that more intelligent subjects *did* show activity in the prefrontal cortex, but only when their tasks were challenging, as in difficult enough for them to have to put some effort into it. This lead to some interesting findings.

Intelligence isn't the work of one dedicated brain region but several, all interlinked. In intelligent people, it seems these links and connections are more efficient and organised, requiring *less* activity overall. Think of it in terms of cars: if you've got a car with an engine roaring like a pack of lions impersonating a hurricane, and a car making no noise whatsoever, the first one isn't automatically going to be the better model. In this case, the noise and activity is because it's trying to do something the more efficient model can do with minimal effort. There's a growing consensus that it's the extent and efficiency of the connections between the regions involved (prefrontal cortex, parietal lobe and so on) that has a big influence on someone's intelligence; the better he or she can communicate and interact, the quicker the processing and the less effort is required to make decisions and calculations.

This is backed up by studies showing that the integrity

and density of white matter in a person's brain is a reliable indicator of intelligence. White matter is the other, often overlooked, kind of tissue in the brain. Grey matter gets all the attention, but 50 per cent of the brain is white matter and it's also very important. It probably gets less publicity because it doesn't 'do' as much. Grey matter is where all the important activity is generated, white matter is made up of bundles and bands of the parts that send the activity to other locations (the axons, the long bit of a typical neuron). If grey matter were the factories, white matter would be the roads needed for delivery and resupply.

The better the white-matter connections between two brain regions, the less energy and effort is required to coordinate them and the tasks they're responsible for, and they're harder to find with a scanner. It's like looking for a needle in a haystack, only instead of a haystack it's a massive pile of slightly bigger needles, and the whole thing is in a washing machine.

Further scanning studies suggest that the thickness of the corpus callosum is also associated with levels of general intelligence. The corpus callosum is the 'bridge' between the left and right hemispheres. It's a big tract of white matter, and the thicker it is the more connections there are between the two hemispheres, enhancing communication. If there's a memory stored on one side that needs to be utilised by the prefrontal cortex on another, a thicker corpus callosum makes this easier and faster. The efficiency and effectiveness of how these regions are connected seems to have a big impact on how well someone can apply their intellect to tasks and problems. As a result of this, brains that are structurally quite different (the size of certain areas, how they're arranged in the cortex, and so on) can display similar levels of intelligence, like two games consoles made by different companies that are similarly powerful.

Now we know efficiency is more important than power. How does that help us go about making ourselves more intelligent? Education and learning is an obvious answer. Actively exposing yourself to more facts, information and concepts means every one you remember will actively increase your crystallised intelligence, and regularly applying your fluid intelligence to as many scenarios as possible will improve matters there. This isn't a cop-out; learning new things and practising new skills can bring about structural changes in the brain. The brain is a plastic organ; it can and will physically adapt to the demands made of it. We met this in Chapter 2: neurons form new synapses when they have to encode a new memory, and this sort of process is found throughout the brain.

For example, the motor cortex, in the parietal lobe, is responsible for planning and control of voluntary movements. Different parts of the motor cortex control different parts of the body, and how much of the motor cortex is dedicated to a body part depends on how much control it needs. Not much of the motor cortex is dedicated to the torso, because you can't do much with it. It's important for breathing and giving your arms somewhere to connect to, but movement-wise we can turn it or bend it slightly, and that's about it. But much of the motor cortex is dedicated to the face and hands, which require a lot of fine control. And that's just for a typical person; studies have revealed that classically trained musicians such as violinists or pianists often have relatively huge areas of the motor cortex dedicated to fine control of the hands and fingers.[9] These people spend all their lives performing increasingly complex and intricate movements with their hands (usually at high speeds), so the brain has adapted to support this behaviour.

Similarly, the hippocampus is needed for spatial memory (memory for places and navigation) as well as episodic memory. This makes sense, given that it is responsible for processing memory for complex combinations of perceptions, which is necessary for navigating your environment. Studies by Professor Eleanor Maguire and her colleagues showed that London taxi drivers with the 'Knowledge' (the required intricate awareness of London's incredibly vast and complicated road network) had an enlarged posterior hippocampus – the navigation part – when compared to non-taxi drivers.[10] These studies were conducted mostly in the days before satnavs and GPS though, so there's no telling how they'd pan out now.

There is even some evidence (although much of it from studies using mice, and how smart can they be?) to suggest that learning new skills and abilities does lead to the white matter involved being enhanced, by increasing the properties of the myelin (the dedicated coating provided by support cells that regulates signal transmission speed and efficiency) around the nerves. So, technically, there are ways to boost your brain power.

That's the good news. Here's the bad.

All of the things mentioned above take much time and effort, and even then the gains can be fairly limited. The brain is complex and responsible for a ridiculous number of functions. As a result, it's easy to increase ability in one region without affecting others. Musicians may have exemplary knowledge of how to read music, listen to cues, dissect sounds and so on, but this doesn't mean they'll be equally good at maths or languages. Enhancing levels of general, fluid intelligence is difficult; it being produced by a range of brain regions and links means it's an especially difficult thing to 'increase' with restricted tasks or methods.

While the brain remains relatively plastic throughout life, much of its arrangement and structure is effectively 'set'. The long white-matter tracts and pathways will have been laid down earlier in life, when development was still under way. By the time we hit our mid-twenties, our brains are essentially fully developed, and it's fine-tuning from thereon in. This is the current consensus anyway. As such, the general view is that fluid intelligence is 'fixed' in adults, and depends largely on genetic and developmental factors during our upbringing (including our parents' attitudes, our social background and education).

This is a pessimistic conclusion for most people, especially those who want a quick fix, an easy answer, a short-cut to enhanced mental abilities. The science of the brain doesn't allow for such things. Sadly but inevitably, there are many people out there who offer them anyway.

Countless companies now sell 'brain-training' games and exercises, which claim to be able to boost intelligence. These are invariably puzzles and challenges of varying difficulty, and it's true that if you play them often enough you will get increasingly better at them. But *only* them. There is, at present, no accepted evidence that any of these products cause an increase in general intelligence; they just cause you to become good at a specific game, and the brain is easily complex enough not to have to enhance everything else to allow this to happen.

Some people, particularly students, have started taking pharmaceuticals such as Ritalin and Adderall, intended to treat conditions like ADHD, when studying for exams, in order to boost concentration and focus. While they might achieve this briefly and in very limited ways, the long-term consequences of taking powerful brain-altering drugs when you don't have the underlying issue they're meant to treat are potentially very worrying. Plus, they can backfire: unnaturally ramping up your focus and concentration with drugs can prove exhausting and depleting to your reserves, meaning you burn out much faster and (for example) sleep through the exam you're studying for.

Drugs meant to improve or enhance mental function are classed as Nootropics, aka 'smart drugs'. Most of these are relatively new and affect only specific processes such as memory or attention, so their long-term effects on general intelligence are currently anyone's guess. The more powerful ones are restricted largely to use in neurodegenerative diseases such as Alzheimer's, where the brain is genuinely degrading at an alarming rate.

There is also a wide variety of foods (for instance, fish oils) that are supposed to increase general intelligence, too, but this is also dubious. They may facilitate one aspect of the brain in one minor way, but this isn't enough for a permanent and widespread boost of intelligence.

There are even technological methods being touted these days, particularly with a technique known as transcranial direct-current stimulation (tCDS). A review by Djamila Bennabi and her colleagues in 2014 found that tCDS (where a low-level current is passed through targeted brain regions) does seemingly enhance abilities such as memory and language in both healthy and mentally ill subjects, and seems to have few to no side-effects thus far. Other reviews and studies have yet to establish a viable effect of the method though. Clearly, there's a lot of work to be done before this sort of thing becomes widely available therapeutically.[11]

Despite this, many companies currently sell gadgets that claim to exploit tCDS for improving performance on things like video games. To avoid libelling anyone, I'm not saying these things don't work, but if they do, that means companies are selling items that actively alter brain activity (as powerful drugs do) via means that aren't scientifically established or understood, to people without any specialist training or supervision. This is a bit like selling antidepressants at the supermarket, next to the chocolate bars and packs of batteries.

So, yes, you can increase your intelligence, but it takes a lot of time and effort over prolonged periods, and you can't just do things you're already good at and/or know. If you get really good at something then your brain becomes so efficient at it, it essentially stops realising it's happening. And if it doesn't know it's happening, it won't adapt or respond to it, so you get a self-limiting effect.

The main problem seems to be that, if you want to be more intelligent, you have to be very determined or very smart in order to outsmart your own brain.

You're pretty smart for a small person

(Why tall people are smarter and the heritability of intelligence)

Tall people are smarter than shorter people. It's true. This is a fact that many find surprising, even offensive (if they're short).

Surely, it's ridiculous to say that someone's height is related to their intelligence? Apparently, it isn't.

Before I get besieged by an enraged but diminutive mob, it's important to point out that this is not an absolute by any means. Basketball players are not automatically more intelligent than jockeys. André the Giant was not smarter than Einstein. Marie Curie would not have been outwitted by Hagrid. The correlation between height and intelligence is usually cited as being about 0.2, meaning height and intelligence seem to be associated in only 1 in 5 people.

Plus, it doesn't make a big difference. Take a random tall person and a random short person and measure their IQs; it's anyone's guess as to who'll be the more intelligent. But you do this often enough, say with 10,000 tall people and 10,000 short people, and the overall pattern will be that the average IQ score of taller people will be slightly higher than that of the shorter people. Might be just 3–4 IQ points' difference, but it's still a pattern, one persistent across numerous studies into the phenomenon.[12] What's going on there? Why would being taller make you more intelligent? It's one of the weird and confusing properties of human intelligence.

One of the more likely causes of this height–intelligence association, according to the available science, is genetic. Intelligence is known to be heritable to a certain degree. To clarify, heritability is the extent to which a property or trait of a person varies due to genetics. Something with a heritability of 1.0 means all possible variation of a trait is due to genes, and a heritability of 0.0 would mean none of the variation is genetic.

For example, your species is purely a result of your genes, so 'species' would have a heritability of 1.0. If your parents were pigs, you'll be a pig, no matter what happens as you grow and develop. There are no environmental factors that will turn a pig into a cow. By contrast, if you are currently on fire, this is purely the result of the environment, so has a heritability of 0.0. There are no genes that cause people to burst into flames; your DNA doesn't cause you to burn constantly and produce little burning babies. However, countless properties of the brain are the result of both genes and environment.

Intelligence itself is heritable to a surprisingly high degree; a review of the available evidence by Thomas J. Bouchard[13] suggests that in adults it's around 0.85, although interestingly it's only about 0.45 in children. This may seem odd; how can genes influence adult intellect more than children's? But this is an inaccurate interpretation of what heritability means. Heritability is a measurement of the extent to which variation among groups is genetic in nature, not the extent to which genes *cause* something. Genes may be just as influential in determining a child's intelligence as an adult's, but with children it seems there are *more* things that can also influence intelligence. Children's brains are still developing and learning, so there's a lot going on that can contribute to apparent intelligence. Adult brains are more 'set'; they've gone through the whole development and maturing process, so external factors aren't so potent any more, so differences between individuals (who in typical societies with compulsory education will have roughly similar learning backgrounds) are more likely to be due to more internal (genetic) differences.

All of this may giving a misleading idea about intelligence and the genes, implying it's a far simpler and more direct arrangement than it is. Some people like to think (or hope) that there is a gene for intelligence, something that could make us smarter if it was activated or strengthened. This seems unlikely; just as intelligence is the sum of many different processes, so these processes are controlled by many different genes, all of which have a part to play. Wondering which gene is responsible for a trait such as intelligence is like wondering which piano key is responsible for a symphony.*

Height is also determined by numerous factors, many of them genetic, and some scientists think that there might be a gene (or genes) that influences intelligence that also influences height, thus providing a link between being tall and being intelligent. It's entirely possible for single genes to have multiple functions. This is known as pleiotropy.

Another argument is that there's no gene(s) that mediate both height and intelligence, but rather the association is due to sexual selection, because both height and intelligence are qualities in men that typically attract women. As a result, tall intelligent men would have the most sexual partners and be more able to spread their DNA through the population via their offspring, all of whom would have the genes for height and intelligence in their DNA.

An interesting theory, but not one that is universally accepted. Firstly, it's very biased towards men, suggesting that they only need to have a couple of attractive traits and women will be inexplicably drawn to them, like moths to a gangly, witty flame. Height is far from the only thing people are attracted to. Also, tall men tend to have taller daughters, and a lot of men are put off and intimidated by tall women (or so my tall female friends tell me).

* Admittedly, there are some genes that are implicated in having a potentially key role in mediating intelligence. For example, the gene apolipoprotein-E, which results in the formation of specific fat-rich molecules with a wide variety of bodily functions, is implicated in Alzheimer's disease and cognition. But the influence of genes on intelligence is breathtakingly complex, even with the limited evidence we currently have, so we won't go into it here.

Same goes for intelligent women (or so my intelligent female friends tell me, which for the record is *all* of them). There's no real actual evidence to suggest that women are invariably attracted to intelligent men either, for various reasons; for instamce, confidence is often considered sexy and, as we've seen, intelligent people can be *less* confident overall. This isn't to mention the fact that intelligence can be unnerving and off-putting; the terms 'nerd' or 'geek' may have been largely reclaimed these days, but they were insults for much of their history, and the stereotype is of them being typically dreadful with the opposite sex. These are just a few examples of how the spread of genes for both height and intelligence could be limited.

Another theory is that growing tall requires access to good health and nutrition, and this may also facilitate brain and therefore intelligence development. It could be as simple as that; greater access to good nutrition and a healthier life during development may result in both increased height and intelligence. It can't be *just* that though, because countless people who have the most privileged and healthy life imaginable end up being short. Or an idiot. Or both.

Could it be to do with brain size? Taller people do have typically bigger brains, and there is a minor correlation between brain size and general intelligence.[14] This is quite a contentious issue. The efficiency of the brain's processing and connections play a big part in an individual's intelligence. but then there is also the fact that certain areas, such as the prefrontal cortex and the hippocampus, are bigger and have more grey matter in people of greater intelligence. Bigger brains would logically make this more likely or possible just by presenting the resources to expand and develop. The general impression seems to be that a bigger brain is maybe yet another contributing factor, but not a definite cause. Big brains perhaps give you more of a chance of becoming intelligent, rather than it being an inevitability? Buying expensive new trainers doesn't actually make you faster at running, but they might encourage you to become so. The same can be said of specific genes, actually.

Genetics, parenting styles, quality of education, cultural norms, stereotyping, general health, personal interests, disorders; all of these and more can lead to the brain being more or less able or likely to perform intelligent actions. You can no more separate human intelligence from human culture than you could separate a fish's development from the water it lives in. Even if you were to separate a fish from the water, its development would only ever be 'brief'.

Culture plays a massive role in how intelligence manifests. A perfect example of this was provided in the 1980s by Michael Cole.[15] He and his team went to the remote Kpelle tribe in Africa, a tribe that was relatively untouched by modern culture and the outside world. They wanted to see if equivalent human intelligence was demonstrated in the Kpelle people, stripped of the cultural factors of Western civilisation. At first, it proved frustrating; the Kpelle people could demonstrate only rudimentary intelligence, and couldn't even solve basic puzzles, the kind a developed-world child would surely have no problem with. Even if the researcher 'accidentally' gave clues as to the right answers, the Kpelle still didn't grasp it. This suggested that their primitive culture wasn't rich or stimulating enough to produce advanced intelligence, or even that some quirk of Kpelle biology prevented them from achieving intellectual sophistication. However, the story is that, frustrated, one of the researchers told them to do the test 'like a fool would', and they immediately produced the 'correct' answers.

Given the language and cultural barriers, the tests involved sorting items into groups. The researchers decided that sorting items into categories (tools, animals, items made of stone, wood, and so on), something that required abstract thinking and processing, was more intelligent. But the Kpelle always sorted things into function (things I can eat, things I can wear, things I can dig with). This was deemed 'less' intelligent, but clearly the Kpelle disagreed. These are people who live off the land, so sorting things into arbitrary categories would be a meaningless and wasteful activity, something a 'fool' would do. As well as being an important lesson in not judging people by your own preconceptions (and maybe about doing better groundwork before beginning an experiment), this example shows how the very concept of intelligence is seriously affected by the environment and preconceptions of society.

A less-drastic example of this is known as the Pygmalion effect. In 1965, Robert Rosenthal and Lenore Jacobson did a study where teachers in elementary schools were told that certain pupils were advanced or intellectually gifted, and should be taught and monitored accordingly.[16] As you'd expect, these pupils showed tests and academic performance in line with being of higher intelligence. The trouble was, they weren't gifted; they were normal pupils. But being treated as if they were smarter and brighter meant they essentially started performing to meet expectations. Similar studies using college students have shown similar results; when students are told that intelligence is fixed, they tend to perform worse on tests. If told that it's variable, they perform better.

Maybe this is another reason why taller people seem more intelligent overall? If you grow taller at a young age, people may treat you as if you're older, so engage you in more mature conversation, so your still-developing brain conforms to these expectations. But in any case, clearly self-belief is important. So any time I've mentioned that intelligence is 'fixed' in this book, I've essentially been hampering your development. Sorry, my bad.

Another interesting/weird thing about intelligence? It's increasing worldwide, and we don't know why. This is called the Flynn effect, and it describes the fact that general scores of intelligence, both fluid and crystallised, are increasing in a wide variety of populations around the world with every generation, in many countries, and despite the varying circumstances that are found in each one. This may be due to improved education worldwide, better healthcare and health awareness, greater access to information and complex technologies, or maybe even the awakening of dormant mutant powers that will slowly turn the human race into a society of geniuses.

There's no evidence to suggest that last one is occurring, but it would make a good film.

There are many possible explanations as to why height and intelligence are linked. They all may be right, or none of them may be right. The truth, as ever, probably lies somewhere between these extremes. It's essentially another example of the classic nature v. nurture argument.

Is it surprising that it would be so uncertain, given what we know about intelligence? It's hard to define, measure and isolate but it's definitely there and we can study it. It is a specific general ability made up of several others. There are numerous brain regions used to produce intelligence, but it may be the manner in which these are connected that makes all the difference. Intelligence is no guarantee of confidence and lack of it is no guarantee of insecurity, because the manner in which the brain works flips the logical arrangement on its head, unless people are treated as if they are intelligent, in which case it seems to make you smarter, so even the brain isn't sure what it's meant to do with the intelligence it is responsible for. And the level of general intelligence is essentially fixed by genes and upbringing, except if you're willing to work at it, in which case it can be increased, maybe.

Studying intelligence is like trying to knit a sweater with no pattern, using candy floss instead of wool. Overall, it's actually incredibly impressive that you can even make the attempt.

5

Did you see this chapter coming?

The haphazard properties of the brain's observational systems

One of the more intriguing and (apparently) uniquely human abilities granted us by our mighty brains is the ability to look 'inwards'. We are self-aware, we can sense our internal state and our own minds, and even assess and study them. As a result, introspection and philosophising are something prized by many. However, how the brain actually perceives the world beyond the skull is also incredibly important, and much of the brain's mechanisms are dedicated to some aspect of this. We perceive the world via our senses, focus on the important elements of it, and act accordingly.

Many may think what we perceive in our heads is a 100 per cent accurate representation of the world as it is, as if the eyes and ears and the rest are essentially passive recording systems, receiving information and passing it on to the brain, which sorts it and organises it and sends it to the relevant places, like a pilot checking the instruments. But that isn't what's happening, at all. Biology is not technology. The actual information that reaches the brain via our senses is not the rich and detailed stream of sights, sounds and sensations that we so often take for granted; in truth, the raw data our senses provide is more like a muddy trickle, and our brain does some quite incredible work to polish it up to give us our comprehensive and lavish world view.

Imagine a police sketch artist, constructing an image of a person from secondhand descriptions. Now imagine it's not one other person who's providing the descriptions, but hundreds. All at once. And it's not a sketch of a person they have to create but a full-colour 3D rendering of the town in which the crime occurred, and everyone in it. And they have to update it every minute. The brain is a bit like that, only probably not quite as harassed as this sketch artist would be.

It is undeniably impressive that the brain can create such a detailed representation of our environment from limited information but errors and mistakes are going to sneak in. The manner in which the brain perceives the world around us, and which parts it deems important enough to warrant attention, is something that illustrates both the awesome power of the human brain, and also its many imperfections.

A rose by any other name . . . (Why smell is more powerful than taste)

As everyone knows, the brain has access to five senses. Although, actually, neuroscientists believe there are more than that.

Several 'extra' senses have been mentioned already, including proprioception (sense of the physical arrangement of body and limbs), balance (the inner-ear-mediated sense that can detect gravity and our movement in space), even appetite, because detecting the nutrient levels in our blood and body is another sort of sense. Most of these are concerned with our internal state, and the five 'proper' ones are responsible for monitoring and perceiving the world around us, our environment. These are, of course, vision, hearing, taste, smell and touch. Or, to be extra scientific, ophthalmoception, audioception, gustaoception, olfacoception and tactioception, respectively (although most scientists don't really use these terms, to save time). Each of these senses is based on sophisticated neurological mechanisms and the brain gets even more sophisticated when using the information they provide. All the senses essentially boil down to detecting things in our environment and translating them into the electrochemical signals used by neurons which are connected to the brain. Coordinating all this is a big job, and the brain spends a lot of time on it.

Volumes could be and have been written about the individual senses, so let's start here with perhaps the weirdest sense, smell. Smell is often overlooked. Literally, what with the nose being right below the eyes. This is unfortunate, as the brain's olfactory system, the bit that smells (as in 'processes odour perception'), is odd and fascinating. Smell is believed to be the first sense to have evolved. It develops very early; it is the first sense to develop in the womb, and it has been shown that a developing baby can actually smell what the mother is smelling. Particles inhaled by the mother end up in the amniotic fluid where the foetus can detect them. It was previously believed that humans could detect up to 10,000 separate odours. Sounds like a lot, but this total was based on a study from the 1920s, which obtained the figure largely from theoretical considerations and assumptions that were never really scrutinised.

Flash forward to 2014, when Caroline Bushdid and her team actually tested this claim, getting subjects to discriminate between chemical cocktails of very similar odours, something that should be practically impossible if our olfactory system is limited to 10,000 smells. Surprisingly, the subjects could do it quite easily. In the end, it was estimated that humans can actually smell in the region of 1 *trillion* odours. This sort of number is usually applied to astronomical distances, not something as humdrum as a human sense. It's like finding out the cupboard where you store the vacuum cleaner actually leads to a subterranean city with a civilisation of mole people.*

So how does smell work? We know smell is conveyed to the brain through the olfactory nerve. There are twelve facial nerves that link the functions of the head to the brain, and the olfactory nerve is number 1 (the optic nerve is number 2). The olfactory neurons that make up the olfactory nerve are unique in many ways, the most pronounced of which is they're one of the few types of human neurons that can regenerate, meaning the olfactory nerve is the Wolverine (of *X-Men* fame) of the nervous system. The regenerative capabilities of these nose neurons means they are extensively studied, with the aim of exploiting their regenerating abilities to apply them to damaged neurons elsewhere – for instance, in the spine of paraplegics.

Olfactory neurons regenerate because they are one of the few types of sensory neurons that are directly exposed to the 'outside' environment, which tends to degrade fragile nerve cells. Olfactory neurons are in the lining of the upper parts of your nose, where the dedicated receptors embedded in them can detect particles. When they come into contact with

* Some scientists have called this finding into question, arguing that this staggering number of smell sensations is more a quirk of questionable maths used in the research than the result of our mighty nostrils.[1]

a specific molecule, they send a signal to the olfactory bulb, the region of the brain responsible for collating and organising information about odour. There are a lot of different smell receptors; a Nobel Prize-winning study by Richard Axel and Linda Buck in 1991 discovered that 3 per cent of the human genome codes for olfactory receptor types.[2] This also supports the idea that human smell is more complex than we'd previously thought.

When the olfactory neurons detect a specific substance (a molecule of cheese, a ketone from something sweet, something emanating from the mouth of someone with questionable dental hygiene) they send electrical signals to the olfactory bulb, which relays this information to areas such as the olfactory nucleus and piriform cortex, meaning you experience a smell.

Smell is very often associated with memory. The olfactory system is located right next to the hippocampus and other primary components of the memory system, so close in fact that early anatomical studies thought that's what the memory system was for. But they're not just two separate areas that happen to be side by side, like an enthusiastic vegan living next to a butcher. The olfactory bulb is part of the limbic system, just like the memory-processing regions, and has active links to the hippocampus and the amygdala. As a result, certain smells are particularly strongly associated with vivid and emotional memories, like how a smell of roast dinner can suddenly remind you of Sundays at your grandparents' house.

You've probably experienced this yourself on many occasions, how a certain smell or odour can trigger powerful memories of childhood and/or bring about emotional moods associated with smells. If you spent a lot of happy time as a child at your grandfather's house and he smoked a pipe, you will likely have a sort of melancholy fondness for the smell of pipe smoke. Smell being part of the limbic system means it has a more direct route to triggering emotions than other senses, which would explain why smell can often elicit a more powerful response than most other senses. Seeing a fresh loaf of bread is a fairly innocuous experience, *smelling* one can be very pleasurable and oddly reassuring, as it's stimulating and coupled with the enjoyable memories of things associated with the smell of baking, which invariably ends up with something pleasant to eat. Smell can have the opposite effect too, of course; seeing rotten meat isn't very nice, but smelling it is what'll make you throw up.

The potency of smell and its tendency to trigger memories and emotions hasn't gone unnoticed. Many try to exploit this for profit: estate agents, supermarkets, candle-makers and more all try to use smell to control people's moods and make them more prone to handing over money. The effectiveness of this approach is known but probably limited by the way in which people vary considerably – someone who's had food poisoning from vanilla ice-cream won't find that odour reassuring or relaxing.

Another interesting misconception about smell: for a long time, it was widely believed that smell can't be 'fooled'. However, several studies have shown this to be not true. People experience illusions of smell all the time, such as thinking a sample smell is pleasant or unpleasant depending on how it's labelled (for instance, 'Christmas tree' or 'toilet cleaner' – and for the record this isn't a joke example; it's a real one from a 2001 experiment by researchers Herz and von Clef).

The reason it was believed there were no olfactory illusions seems to be because the brain only gets 'limited' information from smell. Tests have shown that, with practice, people can 'track' things via their scent, but it's generally restricted to basic detection. You smell something, you know something is nearby that's giving off that smell, and that's about it; it's either 'there' or 'not there'. So if the brain scrambles the smell signals, so that you end up smelling something that's different from what's actually producing the odour, how would you even know? Smell may be powerful, but it's got a limited range of applications for the busy human.

Olfactory hallucinations,* smelling things that aren't there, also exist, and can be worryingly common. People often report the phantom smell of burning – toast, rubber, hair or just a general 'scorched' smell. It's common enough for there to be numerous websites dedicated to it. It's often linked to neurological phenomena, such as epilepsy, tumours or strokes, things that could end up causing unexpected activity in the olfactory bulb or elsewhere in the smell-processing system, and be interpreted as a burning sensation. That's another useful distinction: illusions occur when the sensory system gets something wrong, has been fooled. Hallucinations are more typically an actual malfunction, where something's actually awry in the brain's workings.

Smell doesn't always operate alone. It's often classed as a 'chemical' sense, because it detects and is triggered by

* It's important to clarify the difference between *illusions* and *hallucinations*. Illusions are when the senses detect something but interpret it wrongly, so you end up perceiving something other than what the thing actually is. By contrast, if you smell something *with no source*, this is a hallucination; perceiving something that isn't actually there, which suggests something isn't working as it should deep in the sensory-processing areas of the brain. Illusions are a quirk of the brain's workings; hallucinations are more serious.

specific chemicals. The chemical sense is taste. Taste and smell are often used in conjunction; most of what we eat has a distinct smell. There's also a similar mechanism as receptors in the tongue and other areas of the mouth respond to specific chemicals, usually molecules soluble in water (well, saliva). These receptors are gathered in taste buds, which cover the tongue. It's generally accepted that there are five types of taste bud: salt, sweet, bitter, sour and umami. The last responds to monosodium glutamate, essentially the 'meat' taste. There are actually several more 'types' of taste, such as astringency (for instance from cranberries), pungency (ginger) and metallic (what you get from . . . metal).

Smell is underrated, but taste, by contrast, is a bit rubbish. It is the weakest of our main senses; many studies show taste perception to be largely influenced by other factors. For example, you may be familiar with the practice of wine tasting, where a connoisseur will take a sip of wine and declare that it is a fifty-four-year-old Shiraz from the vineyards of south-west France, with hints of oak, nutmeg, orange and pork (just guessing here) and that the grapes were crushed by a twenty-eight-year-old named Jacques with a verruca on his left heel.

All very impressive and refined, but many studies have revealed that such a precise palate is more to do with the mind than the tongue. Professional wine tasters are typically very inconsistent with their judgements; one professional taster might declare that a certain wine is the greatest ever, while another with identical experience declares it's basically pond water.[3] Surely a good wine will be recognised by everyone? Such is the unreliability of taste that no, it won't. Wine tasters have also been given several samples of wine to taste and been unable to determine which is a celebrated vintage and which is mass-produced cheap slop. Even worse are tests that show wine tasters, given samples of red wine to evaluate, are apparently unable to recognise that they're drinking white wine with food dye in it. So clearly, our sense of taste is no good when it comes to accuracy or precision.

For the record, scientists don't have some sort of bizarre grudge against wine tasters, it's just that there aren't many professions that rely on a well-developed sense of taste to such an extent. And it's not that they're lying; they are almost certainly experiencing the tastes they claim to, but these are mostly the results of expectation, experience and the brain having to get creative, not the actual taste buds. Wine tasters may still object to this constant undermining of their discipline by neuroscientists.

The fact is that tasting something is, in many cases, something of a multisensory experience. People with nasty colds or other nose-clogging maladies often complain about being unable to taste food. Such is the interaction of senses determining taste that they tend to intermingle quite a lot and confuse the brain, and taste, as weak as it is, is constantly being influenced by our other senses, the main one being, you've guessed it, smell. Much of what we taste is derived from the smell of what we're eating. There have been experiments where subjects, with their nose plugged and wearing blindfolds (to rule out vision's influence, too), were unable to discern between apples, potatoes and onions if they had to rely on taste alone.[4]

A 2007 paper by Malika Auvray and Charles Spence[5] revealed that if something has a powerful smell while we're eating it the brain tends to interpret that as a taste, rather than an odour, even if it's the nose relaying the signals. The majority of the sensations are in the mouth, so the brain overgeneralises and assumes that's where everything is coming from and interprets signals accordingly. But the brain already has to do a lot of the work in generating taste sensations, so it would be churlish to begrudge it making inaccurate assumptions.

The take-home message from all of this is that if you're a bad cook, you can still get away with dinner parties if your guests are suffering from terrible head colds and willing to sit in the dark.

Come on, feel the noise
(How hearing and touch are actually related)

Hearing and touch are linked at a fundamental level. This is something most people don't know, but think about it; have you ever noticed how incredibly enjoyable it can be to clean out your ear with a cotton bud? Yes? Well, that's nothing to do with this, I'm just establishing the principle. But the truth is, the brain may perceive touch and hearing completely differently, but the mechanisms it uses to perceive them at all have a surprising amount of overlap.

In the previous section, we looked at smell and taste, and how they often overlap. Admittedly, they do often have similar roles regarding recognising foodstuffs, and can influence each other (smell predominately influencing taste), but the main connection is that smell and taste are both *chemical* senses. The receptors for taste and smell are triggered in the presence of specific chemical substances, like fruit juice or gummy bears.

By contrast, touch and hearing; what do they have in

common? When was the last time you thought something sounded sticky? Or 'felt' high-pitched? Never, right?

Actually, wrong. Fans of the louder types of music often enjoy it at a very tactile level. Consider the sound systems you get in clubs, cars, concerts and so forth that amplify the bass element of music so much that it makes your fillings rattle. When it's powerful enough or of a certain pitch, sound often seems to have a very 'physical' presence.

Hearing and touch are both classed as *mechanical* senses, meaning they are activated by pressure or physical force. This might seem weird, given that hearing is clearly based on sound, but sound is actually vibrations in the air that travel to our eardrum and cause it to vibrate in turn. These vibrations are then transmitted to the cochlea, a spiral-shaped fluid-filled structure, and thus sound travels into our heads. The cochlea is quite ingenious, because it's basically a long, curled-up, fluid-filled tube. Sound travels along it, but the exact layout of the cochlea and the physics of soundwaves mean the frequency of the sound (measured in hertz, Hz) dictates how far along the tube the vibrations travel. Lining this tube is the organ of Corti. It's more of a layer than a separate self-contained structure, and the organ itself is covered with hair cells, which aren't actually hairs, but receptors, because sometimes scientists don't think things are confusing enough on their own.

These hair cells detect the vibrations in the cochlea, and fire off signals in response. But the hair cells only in certain parts of the cochlea are activated due to the specific frequencies travelling only certain distances. This means that there is essentially a frequency 'map' of the cochlea, with the regions at the very start of the cochlea being stimulated by higher-frequency soundwaves (meaning high-pitched noises, like an excited toddler inhaling helium) whereas the very 'end' of the cochlea is activated by the lowest-frequency soundwaves (very deep noises, like a whale singing Barry White songs). The areas between these extremes of the cochlea respond to the rest of the spectrum of sounds audible to humans (between 20 Hz and 20,000 Hz).

The cochlea is innervated by the eighth cranial nerve, named the vestibulocochlear nerve. This relays specific information via signals from the hair cells in the cochlea to the auditory cortex in the brain, which is responsible for processing sound perception, in the upper region of the temporal lobe. And the specific part of the cochlea the signals come from tells the brain what frequency the sound is, so we end up perceiving it as such, hence the cochlea 'map'. Quite clever really.

The trouble is, a system like this, involving a very delicate and precise sensory mechanism essentially being shaken constantly, is obviously going to be a bit fragile. The eardrum itself is made up of three tiny bones arranged in a specific configuration, and this can often be damaged or disrupted by fluid, ear wax, trauma, you name it. The ageing process also means the tissues in the ear get more rigid, restricting vibrations, and no vibrations means no auditory perception. It would be reasonable to say that the gradual age-related decline of the hearing system has as much to do with physics as biology.

Hearing also has a wide selection of errors and hiccups, such as tinnitus and similar conditions, that cause us to perceive sounds that aren't there. These occurrences are known as endaural phenomena; sounds that have no external source, caused by disorders of the hearing system (for example, wax getting into important areas or excessive hardening of important membranes). These are distinct from auditory hallucinations, which are more the result of activity in the 'higher' regions of the brain where the information is processed rather than where it originates. They're usually the sensation of 'hearing voices' (discussed in the later section on psychosis), but other manifestations are musical ear syndrome, where sufferers hear inexplicable music, or the condition where sufferers hear sudden loud bangs or booms, known as exploding head syndrome, which is one from the category 'conditions that sound far worse than they actually are'.

Regardless of this, the human brain still does an impressive job of translating vibrations in the air to the rich and complex auditory sensations we experience every day.

So hearing is a mechanical sense that responds to vibration and physical pressure exerted by sound. Touch is the other mechanical sense. If pressure is applied to the skin, we can feel it. We can do this via dedicated mechanoreceptors that are located everywhere in our skin. The signals from the receptors are then conveyed via dedicated nerves to the spinal cord (unless the stimulation is applied to the head, which is dealt with by the cranial nerves), where they're then relayed to the brain, arriving at the somatosensory cortex in the parietal lobe which makes sense of where the signals come from and allows us to perceive them accordingly. It seems fairly straightforward, so obviously it isn't.

Firstly, what we call touch has several elements that contribute to the overall sensation. As well as physical pressure, there's vibration and temperature, skin stretch and even pain in some circumstances, all of which have their own dedicated receptors in the skin, muscle, organ or bone. All of this is known as the somatosensory system (hence somatosensory cortex) and our whole body is innervated by the nerves that serve it. Pain, aka nociception, has its own dedicated receptors and nerve fibres throughout the body.

Pretty much the only organ that doesn't have pain receptors is the brain itself, and that's because it's responsible for receiving and processing the signals. You could argue that the brain feeling pain would be confusing, like trying to call your own number from your own phone and expecting someone to pick up.

What is interesting is that touch sensitivity isn't uniform; different parts of the body respond differently to the same contact. Like the motor cortex discussed in a previous chapter, the somatosensory cortex is laid out like a map of the body corresponding to the areas it's receiving information from, with the foot region processing stimuli from feet, the arm region for the arm, and so on.

However, it doesn't use the same dimensions as the actual body. This means that the sensory information received doesn't necessarily correspond with the size of the region the sensations are coming from. The chest and back areas take up quite a small amount of space in the somatosensory cortex, whereas the hands and lips take up a very large area. Some parts of the body are far more sensitive to touch than others; the soles of the feet aren't especially sensitive, which makes sense as it wouldn't be practical to feel exquisite pain whenever you step on a pebble or a twig. But the hands and lips occupy disproportionately large areas of the somatosensory cortex because we use them for very fine manipulation and sensations. Consequently, they are very sensitive. As are the genitals, but let's not go into that.

Scientists measure this sensitivity by simply prodding someone with a two-pronged instrument and seeing how close together these prongs can be and still be recognised as separate pressure points.[6] The fingertips are especially sensitive, which is why braille was developed. However, there are some limitations: braille is a series of separate specific bumps because the fingertips aren't sensitive enough to recognise the letters of the alphabet when they're text sized.[7]

Like hearing, the sense of touch can also be 'fooled'. Part of our ability to identify things with touch is via the brain being aware of the arrangement of your fingers, so if you touch something small (for instance, a marble) with your index and middle finger, you'll feel just the one object. But if you cross your fingers and close your eyes, it feels more like two separate objects. There's been no direct communication between the touch-processing somatosensory cortex and the finger-moving motor cortex to flag up this point up, and the eyes are closed so aren't able to provide any information to override the inaccurate conclusion of the brain. This is the Aristotle illusion.

So there are more overlaps between touch and hearing than is immediately apparent, and recent studies have found evidence that the link between the two may be far more fundamental than previously thought. While we've always understood that certain genes were strongly linked to hearing abilities and increased risk of deafness, a 2012 study by Henning Frenzel and his team[8] discovered that genes also influenced touch sensitivity, and interestingly that those with very sensitive hearing also showed a finer sense of touch too. Similarly, those with genes that resulted in poor hearing also had a much higher likelihood of showing poor touch sensitivity. A mutated gene was also discovered that resulted in both impaired hearing and touch.

While there is still more work to be done on this area, this does strongly suggest that the human brain uses similar mechanisms to process both hearing and touch, so deep-seated issues that affect one can end up affecting the other. This is perhaps not the most logical arrangement, but it's reasonably consistent with the taste–smell interaction we saw in the previous section. The brain does tend to group our senses together more often than seems practical. But on the other hand, it does suggest people can 'feel the rhythm' in a more literal manner than is generally assumed.

Jesus has returned . . . as a piece of toast?
(What you didn't know about the visual system)

What do toast, tacos, pizza, ice-cream, jars of spread, bananas, pretzels, crisps and nachos have in common? The image of Jesus has been found in all of them (seriously, look it up). It's not always food though; Jesus often pops up in varnished wooden items. And it's not always Jesus; sometimes it's the Virgin Mary. Or Elvis Presley.

What's actually happening is that there are uncountable billions of objects in the world that have random patterns of colour or patches that are either light or dark, and by sheer chance these patterns sometimes resemble a well-known image or face. And if the face is that of a celebrated figure with metaphysical properties (Elvis falls into this category for many) then the image will have more resonance and get a lot of attention.

The weird part (scientifically speaking) is that even those who are aware that it's just a grilled snack and not the bread-based rebirth of the Messiah can still *see* it. Everyone can still recognise what is said to be there, even if they dispute the origins of it.

The human brain prioritises vision over all other senses, and the visual system boasts an impressive array of oddities. As with the other senses, the idea that the eyes capture everything about our outside world and relay this information intact to the brain like two worryingly squishy video cameras is a far cry from how things really work.*

Many neuroscientists argue that the retina *is* part of the brain, as it develops from the same tissue and is directly linked to it. The eyes take in light through the pupils and lenses at the front, which lands on the retina at the back. The retina is a complex layer of photoreceptors, specialised neurons for detecting light, some of which can be activated by as little as half-a-dozen photons (the individual 'bits' of light). This is very impressive sensitivity, like a bank security system being triggered because someone had a thought about robbing the place. The photoreceptors that demonstrate such sensitivity are used primarily for seeing contrasts, light and dark, and are known as rods. These work in low-light conditions, such as at night. Bright daylight

* Not that the eyes aren't impressive, because they are. The eyes are so complex that they're often cited (not a pun) by creationists and others opposed to evolution as clear proof that natural selection isn't real; the eye is so intricate it couldn't just 'happen' and therefore must be the work of a powerful creator. But if you truly look at the workings of the eye, then this creator must have designed the eye on a Friday afternoon, or while hung over on the morning shift, because a lot of it doesn't make much sense.

actually oversaturates them, rendering them useless; it's like trying to pour a gallon of water into an egg cup. The other (daylight-friendly) photoreceptors detect photons of certain wavelengths, which is how we perceive colour. These are known as cones, and they give us a far more detailed view of the environment, but they require a lot more light to be activated, which is why we don't see colours at low light levels.

Photoreceptors aren't spread uniformly across the retina. Some areas have different concentrations from others. We have one area in the centre of the retina that recognises fine detail, while much of the periphery gives only blurry outlines. This is due to the concentrations and connections of the photoreceptor types in these areas. Each photoreceptor is connected to other cells (a bipolar cell and a ganglion cell usually), which transmit the information from the photoreceptors to the brain. Each photoreceptor is part of a receptive field (which is made up of all the receptors connected to the same transmission cells) that covers a specific part of the retina. Think of it like a mobile-phone mast, which receives all the different information relayed from the phones within its coverage range and processes them. The bipolar and ganglion cells are the mast, the receptors are the phones; thus there is a specific receptive field. If light hits this field it will activate a specific bipolar or ganglion cell via the photoreceptors attached to it, and the brain recognises this.

In the periphery of the retina, the receptive fields can be quite big, like a golf umbrella canvas around the central shaft. But this means precision suffers – it's difficult to work out where a raindrop is falling on a golf umbrella; you just know it's there. Luckily, towards the centre of the retina, the receptive fields are small and dense enough to provide sharp and precise images, enough for us to be able to see very fine details like small print.

Bizarrely, only one part of the retina is able to recognise this fine detail. It is named the fovea, in the dead centre of the retina, and it makes up less than 1 per cent of the total retina. If the retina were a widescreen TV, the fovea would be a thumbprint in the middle. The rest of the eye gives us more blurry outlines, vague shapes and colours.

You may think this makes no sense, because surely people see the world crisp and clear, give or take the odd cataract? This described arrangement would be more like looking through the wrong end of a telescope made of Vaseline. But, worryingly, that is what we 'see', in the purest sense. It's just that the brain does a sterling job of cleaning this image up before we consciously perceive it. The most convincing Photoshopped image is little more than a crude sketch in yellow crayon compared to the polishing the brain does with our visual information. But how does it do this?

The eyes move around a lot, and much of this is due to the fovea being pointed at various things in our environment that we need to look at. In the old days, experiments tracking eyeball movements used specialised *metal* contact lenses. Just let that sink in, and appreciate how committed some people are to science.*

Essentially, whatever we're looking at, the fovea scans as much of it as possible, as quickly as possible. Think of a spotlight aimed at a football field operated by someone in the

* Modern camera and computing technology means it's much easier (and considerably less uncomfortable) to track eye movements. Some marketing companies have even used eye scanners mounted on trolleys to observe what customers are looking at in shops. Before this, head-mounted laser trackers were used. Science is so advanced these days that lasers are now old-fashioned. This is a cool thing to realise.

middle of a near-lethal caffeine overdose, and you're sort of there. The visual information obtained via this process, coupled with the less-detailed but still-usable image of the rest of the retina, is enough for the brain to do some serious polishing and make a few 'educated guesses' about what things look like, and we see what we see.

This seems a very inefficient system, relying on such a small area of retina to do so much. But considering how much of the brain is required to process this much visual information, even doubling the size of the fovea so it's more than 1 per cent of the retina would require an increase in brain matter for visual processing to the point where our brains could end up the size of basketballs.

But what of this processing? How does the brain render such detailed perception from such crude information? Well, photoreceptors convert light information to neuronal signals which are sent to the brain along the optic nerves (one from each eye).* The optic nerve relays visual information to several parts of the brain. Initially, the visual information is sent to the thalamus, the old central station of the brain, and from there it's spread far and wide. Some of it ends up in the brainstem, either in a spot called the pretectum, which dilates or contracts pupils in response to light intensity, or in the superior colliculus, which controls movement of the eyes in short jumps called saccades.

* For the record, some people claim that they've had eye surgery and their eye was 'taken out' and left dangling on their cheek at the end of the optic nerve, like in a Tex Avery cartoon. This is impossible; there is some give in the optic nerve, but certainly not enough to support the eye like a grotesque conker on a string. Eye surgery usually involves pulling the eyelids back, holding the eye in place with clamps, and numbing injections, so it feels weird from the patient's perspective. But the firmness of the eye socket and fragility of the optic nerve means popping the eye out would effectively destroy it, which isn't a great move for an ophthalmic surgeon.

If you concentrate on how your eyes move when you look from right to left or vice versa, you will notice that they don't move in one smooth sweep but a series of short jerks (do it slowly to appreciate this properly). These movements are saccades, and they allow the brain to perceive a continuous image by piecing together a rapid series of 'still' images, which is what appears on the retina between each jerk. Technically, we don't actually 'see' much of what's happening between each jerk, but it's so quick we don't really notice, like the gap between the frames of an animation. (The saccade is one of the quickest movements the human body can make, along with blinking and closing a laptop as your mum walks into your bedroom unexpectedly.)

We experience the jerky saccades whenever we move our eyes from one object to another, but if we're visually following something in motion our eye movement is as smooth as a waxed bowling ball. This makes evolutionary sense; if you're tracking a moving object in nature it's usually prey or a threat, so you'd need to keep focused on it constantly. But we can do it only when there's something moving that we can track. Once this object leaves our field of vision, our eyes jerk right back to where they were via saccades, a process termed the Optokinetic reflex. Overall, it means the brain *can* move our eyes smoothly, it just often doesn't.

But why when we move our eyes do we not perceive the world around us as moving? After all, it all looks the same as far as images on the retina are concerned. Luckily, the brain has a quite ingenious system for dealing with this issue. The eye muscles receive regular inputs from the balance and motion systems in our ears, and use these to differentiate between eye motion and motion in or of the world around us. It means we can also maintain focus on an object when we're in motion. It's a system that can be confused though, as the motion-detection systems can sometimes end up sending signals to the eyes when we're not moving, resulting in involuntary eye movements called nystagmus. Health professionals look out for these when assessing the health of the visual system, because when your eyes are twitching for no reason, that's not great. It's suggestive of something gone awry in the fundamental systems that control your eyes. Nystagmus is to doctors and optometrists what a rattling in the engine is to a mechanic; might be something fairly harmless, or it might not, but either way it's *not meant to be happening*.

This is what your brain does just working out where to point the eyes. We haven't even started on how the visual information is processed.

Visual information is mostly relayed to the visual cortex in the occipital lobe, at the back of the brain. Have you ever experienced the phenomenon of hitting your head and 'seeing stars'? One explanation for this is that impact causes your brain to rattle around in your skull like a hideous bluebottle trapped in an egg cup, so the back of your brain bounces off your skull. This causes pressure and trauma to the visual processing areas, briefly scrambling them, and as a result we see sudden weird colours and images resembling stars, for want of a better description.

The visual cortex itself is divided into several different layers, which are themselves often subdivided into further layers.

The primary visual cortex, the first place the information from the eyes arrives in, is arranged in neat 'columns', like sliced bread. These columns are very sensitive to orientation, meaning they respond only to the sight of lines of a certain direction. In practical terms, this means we recognise edges. The importance of this can't be overstressed: edges mean boundaries, which means we can recognise individual objects and focus on them, rather than on the uniform surface that makes up much of their form. And it means we can track their movements as different columns fire in response to changes. We can recognise individual objects and their movement, and dodge an oncoming football, rather than just wonder why the white blob is getting bigger. The discovery of this orientation sensitivity is so integral that when David Hubel and Torsten Wiesel discovered it in 1981, they ended up with a Nobel Prize.[9]

The secondary visual cortex is responsible for recognising colours, and is extra impressive because it can work out colour constancy. A red object in bright light will look, on the retina, very different from a red object in dark light, but the secondary visual cortex can seemingly take the amount of light into account, and work out what colour the object is 'meant' to be. This is great, but it's not 100 per cent reliable. If you've ever argued with someone over what colour something is (such as whether a car is dark blue or black) you've experienced first hand what happens when the secondary visual cortex gets confused.

It goes on like this, the visual-processing areas spreading out further into the brain, and the further they spread from the primary visual cortex the more specific they get regarding what it is they process. It even crosses over into other lobes, such as the parietal lobe containing areas that process spatial awareness, to the inferior temporal lobe processing recognition of specific objects and (going back to the start) faces. We have parts of the brain that are dedicated to recognising faces, so we see them everywhere. Even if they're not there, because it's just a piece of toast.

These are just some of the impressive facets of the visual system. But perhaps the one that is most fundamental is the fact that we can see in three dimensions, or '3D' as the kids are calling it. It's a big ask, because the brain has to create a rich 3D impression of the environment from a patchy 2D image. The retina itself is technically a 'flat' surface, so it can't support 3D images any more than a blackboard can. Luckily, the brain has a few tricks to get around this.

Firstly, having two eyes helps. They may be close together on the face, but they're far enough apart to supply subtly different images to the brain, and the brain uses this difference to work out depth and distance in the final image we end up perceiving.

It doesn't just rely on the parallax resulting from ocular disparity (that's the technical way of saying what I just said) though, as this requires two eyes to be working in unison, but when you close or cover one eye, the world doesn't instantly convert to a flat image. This is because the brain can also use aspects of the image delivered by the retina to work out depth and distance. Things like occlusion (objects covering other objects), texture (fine details in a surface if it's close but not if it's far away) and convergence (things up close tend to be much further apart than things in the distance; imagine a long road receding to a single point) and more. While having two eyes is the most beneficial and effective way to work out depth, the brain can get by fine with just one, and can even keep performing tasks that involve fine manipulation. I once knew a successful dentist who could see out of only one eye; if you can't manage depth perception, you don't last long in that job.

These visual-system methods of recognising depth are exploited by 3D films. When you look at a movie screen, you can see the necessary depth because all the required cues discussed above are there. But to a certain extent you are still aware that you're looking at images on a flat screen, because that is the case. But 3D films are essentially two slightly different streams of images on top of each other. Wearing 3D glasses filters out these images, but one lens filters out a specific image and the other filters out the other. As a result, each eye receives a subtly different image. The brain recognises this as depth, and suddenly images on the screen leap out at us and we have to pay double the price for a ticket.

Such is the complexity and density of the visual-system processing that there are many ways it can be fooled. The Jesus-in-a-piece-of-toast phenomenon occurs because there is a temporal-cortex region of the visual system responsible for recognising and processing faces, so anything that looks a bit like a face will be perceived as a face. The memory system can chip in and say if it's a familiar face or not, too. Another common illusion makes two things that are exactly the same colour look different when placed on different backgrounds. This can be traced to the secondary visual cortex getting confused.

Other visual illusions are more subtle. The classic 'is it two faces looking at each other or actually a candlestick?' image is possibly the most familiar. This image presents two possible interpretations, both images are 'correct' but are mutually exclusive. The brain really doesn't handle ambiguity well, so it effectively imposes order on what it's receiving by picking one possible interpretation. But it can change its mind, too, as there are two solutions.

All this barely scratches the surface. It's not really possible to convey the true complexity and sophistication of the visual-processing system in a few pages, but I felt it worth the attempt because vision is so complex a neurological process that underpins so much of our lives, and most people think nothing of it until it starts going awry. Consider this section just the tip of the iceberg of the brain's visual system; there's a vast amount more in the depths below it. And you can perceive such depths only because the visual system is as complex as it is.

Why your ears are burning
(Strengths and weaknesses of human attention, and why you can't help eavesdropping)

Our senses provide copious information but the brain, despite its best efforts, cannot deal with all of it. And why should it? How much is actually relevant? The brain is an incredibly demanding organ in terms of resources, and using it to focus intently on a patch of drying paint would just squander them. The brain *has* to pick and choose what gets noticed. As such, the brain is able to direct perception and conscious processing to things of potential interest. This is attention, and how we use it plays a big role in what we observe of the world around us. Or, often more importantly, what we fail to observe.

For the study of attention, there are two important questions. One is, what's the brain's capacity for attention? How much can it realistically take in before it gets overwhelmed?

The other is, what is it that determines where the attention is directed? If the brain is constantly being bombarded with sensory information, what is it about certain stimuli or input that prioritises it over other things?

Let's start with capacity. Most people have noticed attention has a limited capacity. You've probably experienced a group of people all trying to talk to you at once, 'clamouring for attention'. This is frustrating, usually resulting in loss of patience and shouts of, 'One at a *time*!'

Early experiments, such as those by Colin Cherry in 1953,[10] suggested attention capacity was alarmingly limited, demonstrated by a technique called 'dichotic listening'. This is where subjects wear headphones and receive a different audio stream (typically, a sequence of words) in each ear. They were told they had to repeat the words received in one ear, but then were asked what they could recall from the other ear. Most could identify whether the voice was male or female, but that's it, not even what language was spoken. So attention has such a limited capacity, it can't be stretched beyond a single audio stream.

These and similar findings resulted in 'bottleneck' models of attention, which argued that all the sensory information that is presented to the brain is filtered through the narrow space offered by attention. Think of a telescope: it provides a very detailed image of a small part of the landscape or sky. But, beyond that, there's nothing.

Later experiments changed things. Von Wright and his colleagues in 1975 conditioned subjects to expect a shock when they heard certain words. Then they did the dichotic-listening task. The stream in the *other* ear, not the focus of attention, featured the shock-provoking words. Subjects still showed a measurable fear reaction when the words were heard, revealing that the brain was clearly paying attention to the 'other' stream. But it doesn't reach the level of *conscious* processing, so we aren't aware of it. The bottleneck models break down in the face of data like this, showing people can still recognise and process things 'outside' of the supposed boundaries of attention.

This can be demonstrated in less clinical surroundings. The title of this section refers to when people say their 'ears are burning'. The phrase usually used to mean someone has overheard others talking about them. It occurs often, particularly at a social occasions such as wedding receptions, leaving parties, sporting events, where a lot of people are gathered in various groups, all talking at once. At some point, you'll be having a perfectly enjoyable conversation about your mutual interests (football, baking, celery, whatever), when someone within earshot says your name. They aren't part of your current group; maybe you didn't even know they were there. But they said your name, perhaps followed by the words, 'is a tremendous waste of skin', and suddenly you're paying attention to their conversation, rather than the one you are having, wondering why you ever asked that person to be your best man.

If attention was as limited as the bottleneck models suggest, then this should be impossible. But, clearly, it isn't. This occurrence is known as 'the cocktail-party effect', because professional psychologists are a refined bunch.

The limitations of the bottleneck model lead to formation of the capacity model, typically attributed to work by Daniel Kahneman in 1973,[11] but expounded on by many since. Whereas bottleneck models argued that there is one 'stream' of attention that hops about like a spotlight depending on where it's needed, the capacity model argues that attention is more like a finite resource that can be divided between multiple streams (focuses of attention) so long as the resources are not exhausted.

Both proposed models explain why multitasking is so difficult; with bottleneck models, you have one single stream of attention that keeps leaping between different tasks, making it very difficult to keep track. The capacity model would allow you to pay attention to more than one thing at a time, but only so far as you have the resources to process them effectively; as soon as you go beyond your capacity, you lose the ability to keep track of what's going on. And the resources are limited enough to make it look like a 'single' stream is all we've got in many scenarios.

But *why* this limited capacity? One explanation is that attention is strongly associated with working memory, what we use to store the information we're consciously processing. Attention provides the information to be processed, so if working memory is already 'full', adding more information is going to be difficult, if not impossible. And we know working (short-term) memory has a limited capacity.

This is often sufficient for your typical human, but context is crucial. Many studies focus on how attention is used while driving, where a lack of attention can have serious consequences. In the UK, driving while physically using a phone is not allowed; you have to use a hands-free set-up and keep both hands on the wheel. But a study from the University of Utah in 2013 revealed that, in terms of how it affects performance, using a hands-free set-up is just as bad as using the phone with your hands, because both require a similar amount of attention.[12]

The fact that you have two hands on the wheel as opposed to one may provide some advantage, but the study measured overall speed of responses, scanning of environment, noticing important cues; all these and more are reduced to a similar worrying extent whether using hands-free or not, because they require similar levels of attention. You may well be keeping your eyes on the road, but that's irrelevant if you're ignoring what your eyes are showing you.

Even more worrying, the data suggests it's not just the phone: changing the radio or carrying on a conversation with a passenger can also be equally distracting. With increased technology found in cars and on phones (it's technically not illegal at present to check your emails while driving) the options for distraction are bound to increase.

With all this, you may wonder how anyone can drive for more than ten minutes straight without ending up in a disastrous wreck. It's because we're talking about *conscious* attention, which is where the capacity is limited. As we've discussed, do something often enough and the brain adapts to it, allowing procedural memory, described in Chapter 2. People say they can do something 'without thinking', and that's quite accurate here. Driving can be an anxious, overwhelming experience for beginners, but eventually it becomes so familiar the unconscious systems take over, so conscious attention can be applied elsewhere. However, driving is not something that can be done entirely without thinking; taking account of all other road users and hazards needs conscious awareness, as these are different each time.

Neurologically, attention is supported by many regions, one of which is that repeat offender the prefrontal cortex, which makes sense as that's where working memory is processed. Also implicated is the anterior cingulate gyrus, a large and complex region deep in the temporal lobe that also extends into the parietal lobe, where a lot of sensory information is processed and linked to higher functions such as consciousness.

But the attention controlling systems are quite diffuse, and this has consequences. In Chapter 1, we saw how more advanced conscious parts of the brain and the more primitive 'reptile' elements often end up getting in each other's way. The attention-controlling systems are similar; better organised, but a familiar combination or conflict of conscious and subconscious processing.

For example, attention is directed by exogenous and endogenous cues. Or, in plain English, it has both bottom-up and top-down control systems. Or, even more simply, our attention responds to stuff that happens either outside our head, or inside it. Both of these are demonstrated by the cocktail-party effect, where we direct our attention to specific sounds, also known as 'selective listening'. The sound of your name suddenly causes your attention to shift to it. You didn't know it was coming; you weren't consciously aware of it until it had happened. But, once aware of it, you direct your attention to the source, excluding anything else. An external sound diverted your attention, demonstrating a bottom-up attention process, and your conscious desire to hear more keeps your attention there, demonstrating an internal top-down attention process originating in the conscious brain.*

* Exactly how we 'focus' aural attention is unclear. We don't swivel our ears towards interesting sounds. One possibility comes from a study by Edward Chang and Nima Mesgarani of the University of California, San Francisco, who looked at the auditory cortex of three epilepsy patients who had electrodes implanted in the relevant regions (to record and help localise seizure activity, not for fun or

However, most attention research focuses on the visual system. We can and do physically point our eyes at the subject of attention, and the brain relies mostly on visual data. It's an obvious target for research, and this research has produced a lot of information about how attention works.

The frontal eye fields, in the frontal lobe, receive information from the retinas and create a 'map' of the visual field based on this, supported and reinforced by more spatial mapping and information via the parietal lobe. If something of interest occurs in the visual field, this system can very quickly point the eyes in that direction, to see what it is. This is called overt or 'goal' orientation, as your brain has a goal that is 'I want to look at that!' Say you see a sign that reads SPECIAL OFFER: FREE BACON, then you direct your attention to it straight away, to see what the deal is, to complete the goal of getting bacon. The conscious brain drives the attention, so it's a top-down system. Alongside all this there's another system at work, called *covert* orientation, which is more of a 'bottom-up' one. This system means something is detected that is of biological significance (for instance, the sound of a tiger growling nearby, or a crack from the tree branch your standing on) and attention is automatically directed towards it, *before* the conscious areas of the brain even know what's going on, hence it's a bottom-up system. This system uses the same visual input as the other one as well as sound cues, but is supported by a different set of neural processes in different regions.

---

anything).[13] When asked to focus on a specific audio stream out of two or more heard at once, only the one being paid attention to produced any activity in the auditory cortex. The brain somehow suppresses any competing information, allowing full attention to be paid to the voice being listened to. This suggests your brain really can 'tune someone out', like when they won't stop droning on about their tedious hedgehog-spotting hobby.

According to current evidence, the most widely supported model is one where, on detection of a something potentially important, the posterior parietal cortex (already mentioned regarding vision processing) disengages the conscious attention system from whatever it's currently doing, like a parent switching the television off when their child is meant to put the bins out. The superior colliculus in the midbrain then moves the attention system to the desired area, like a parent moving their child to the kitchen where the bins are. The pulvinar nucleus, part of the thalamus, then reactivates the attention system, like a parent putting bin bags in their child's hand and pushing the child towards the door to put the damn things out!

This system can overrule the conscious, goal-orientated top-down system, which makes sense as it's something of a survival instinct. The unfamiliar shape in your vision could turn out to be an oncoming attacker, or that boring office colleague who insists on talking about his athlete's foot.

These visual details don't have to appear in the fovea, the important middle bit of the retina, to attract our attention. Visually paying attention to something typically involves moving the eyes, but *it doesn't have to.* You'll have heard of 'peripheral vision', where you see something you're not looking at directly. It won't be greatly detailed, but if you're at your desk working at your computer and see an unexpected movement in the corner of your vision that seems the right size and location to be a large spider, you maybe don't want to look at it, in case that's exactly what it is. While you carry on typing, you're very alert to any movement in that particular spot, just waiting to see it again (while hoping not to). This shows that the focus of attention isn't tied directly to where the eyes are pointing.

As with the auditory cortex the brain can specify which part of the visual field to focus on, and the eyes don't have to move to allow it. It may sound like the bottom-up processes are the most dominant, but there's more to it. Stimulus orientation overrides the attention system when it detects a significant stimulus, but it's often the conscious brain that determines what's 'significant' by deciding the context. A loud explosion in the sky would certainly be something that would count as significant, but, if you're going for a walk on 5 November (or 4 July for Americans), an *absence* of explosions in the sky would be more significant, as the brain is expecting fireworks.

Michael Posner, one of the dominant figures in the field of attention research, devised tests that involve getting subjects to spot a target on screen that is preceded by cues which may or may not predict the target location. If there are as few as two cues to look at, people tend to struggle. Attention can be divided between two different modalities (doing a visual test and a listening test at the same time) but if it's anything more complex than a basic yes/no detection test, people typically fall apart trying it. Some people can do two simultaneous tasks if one is something they're very adept at, such as an expert typist doing a maths problem while typing. Or, to use an earlier example, an experienced driver holding a detailed conversation while operating a vehicle.

Attention can be very powerful. One well-known study concerned volunteers at Uppsala University in Sweden,[14] where subjects reacted with sweaty palms to images of snakes and spiders that were on shown on screen for less than 1/300th of a second. It usually takes about half a second for the brain to process a visual stimulus sufficiently for us to consciously recognise it, so subjects were experiencing responses to pictures of spiders and snakes in less than a tenth of the time it actually takes to 'see' them. We've already established that the unconscious attention system responds to biologically relevant cues, and that the brain is primed to spot anything that might be dangerous and has seemingly evolved a tendency to fear natural threats like our eight-legged or no-legged friends. This experiment is a great demonstration of how attention spots something and rapidly alerts the parts of the brain that mediate responses before the conscious mind has even finished saying, 'Huh? What?'

In other contexts, attention can miss important and very unsubtle things. As with the car example, too much occupying our attention means we miss very important things, such as pedestrians (or, more importantly, fail to miss them). A stark example of this was provided by Dan Simons and Daniel Levin in 1998.[15] In their study, an experimenter approached random pedestrians with a map and asked them directions. While the pedestrians were looking at the map, a person carrying a door walked between them and the experimenter. In the brief moment when the door presented an obstruction, the experimenter changed places with someone who didn't look or sound anything like the original person. At least 50 per cent of the time, the map-consulting person didn't notice *any* change, even though they were talking to a different person from the one they'd been speaking to *seconds earlier*. This invokes a process known as 'change blindness', where our brains are seemingly unable to track an important change in our visual scene if it's interrupted even briefly.

This study is known as the 'door study', because the door is the most interesting element here, apparently. Scientists are a weird bunch.

The limits of human attention can and do have serious scientific and technological consequences too. For example, heads-up displays, where the instrument display in machines such as aeroplanes and space vehicles is projected onto the screen or canopy rather than read-outs in the cockpit area, seemed like a great idea for pilots. It saves them having to look down to see their instruments, thus taking their eyes off what's going on outside. Safer all round, right?

No, not really. It turned out when a heads-up display is even slightly too cluttered with information, the pilot's attention is maxed out.[16] They can see right through the display, but they're not *looking* through it. Pilots have been known to land their plane on top of another plane as a result of this (in simulations, thankfully). NASA itself has spent a lot of time working out the best ways to make heads-up displays workable, at the expense of hundreds of millions of dollars.

These are just some of the ways the human attention system can be seriously limited. You might like to argue otherwise, but if you do you clearly haven't been paying attention. Luckily, we've now established you can't really be blamed for that.

6

Personality: a testing concept

The complex and confusing properties of personality

Personality. Everybody has one (except maybe those who enter politics). But what is a personality? Roughly, it's a combination of an individual's tendencies, beliefs, ways of thinking and behaving. It's clearly some 'higher' function, a combination of all the sophisticated and advanced mental processes humans seem uniquely capable of thanks to our gargantuan brains. But, surprisingly, many think personality doesn't come from the brain at all.

Historically, people believed in dualism; the idea that the mind and body are separate. The brain, whatever you think of it, is still part of the body; it's a physical organ. Dualists would argue that the more intangible, philosophical elements of a person (beliefs, attitudes, loves and hates) are held within the mind, or 'spirit', or whatever term is given to the immaterial elements of a person.

Then, on 13 September 1848, as a result of an unplanned explosion, railroad worker Phineas Gage had his brain impaled by a metre-long iron rod. It entered his skull just under his left eye, passed right *through* his left frontal lobe, and exited via the top of his skull. It landed some 25 metres away. The force propelling the rod was so great that a human head offered as much resistance as a net curtain. To clarify, this was not a paper cut.

You'd be forgiven for assuming this would have been fatal. Even today, 'huge iron rod right through the head' sounds like a 100-per-cent-lethal injury. And this happened in the mid-1800s, when stubbing your toe usually meant a grim death from gangrene. But, no, Gage survived, and lived another twelve years.

Part of the explanation for this is that the iron pole was very smooth and pointed, and travelling at such a speed that the wound was surprisingly precise and 'clean'. It destroyed almost all the frontal lobe in the left hemisphere of his brain but the brain has impressive levels of redundancy built into it, so the other hemisphere picked up the slack and provided normal functioning. Gage has become iconic in the fields of psychology and neuroscience, as his injury supposedly resulted in a sudden and drastic change in his personality. From a mild-mannered and hardworking sort, he became irresponsible, ill-tempered, foul-mouthed, and even psychotic. 'Dualism' had a fight on its hands as this discovery firmly established the idea that the workings of the brain are responsible for a person's personality.

However, reports of Gage's changes vary wildly, and towards the end of his life, he was employed long-term as a stagecoach driver, a job with a lot of responsibility and public interaction, so even if he did experience disruptive personality changes he must have got better again. But the extreme claims persist, largely because contemporary psychologists (at the time, a career dominated by self-aggrandising wealthy white men, whereas now it's . . . actually, never mind) leapt on Gage's case as an opportunity to promote their own theories about how the brain worked; and if that meant attributing things that never happened to a lowly railway worker, what of it? This was the nineteenth century, he wasn't exactly going to find out via Facebook. Most of the extreme claims about his personality changes were seemingly made after his death, so it was practically impossible to refute them.

But even if people were dedicated enough to investigate the actual personality or intellectual changes Gage had experienced, how would they do this? IQ tests were half a century away, and that's just one possible property that might have been affected. So Gage's case lead to two persistent realisations about personality: it's a product of the brain, and it's a real pain to measure in a valid, objective manner.

E. Jerry Phares and William Chaplin, in their 2009 book *Introduction to Personality*,[1] came up with a definition of personality that most psychologists would be willing to accept: 'Personality is that pattern of characteristic thoughts, feelings, and behaviours that distinguishes one person from another and that persists over time and situations.'

In the next few sections, we're going to look at a few fascinating aspects – the approaches used to measure personality, what it is that makes people angry, how they end up compelled to do certain things, and that universal arbiter of a good personality, sense of humour.

Nothing personal
(The questionable use of personality tests)

My sister Katie was born when I was three, when my own puny brain was still relatively fresh. We had the same parents, grew up at the same time, in the same place. It was the 1980s in a small isolated Welsh valley community. Overall, we had very similar environments, and very similar DNA.

You might expect us to have very similar personalities. This is the opposite of what happened. My sister was, to put it mildly, a hyperactive nightmare, whereas I was typically so placid you had to poke me to make sure I was conscious. We're both adults now, and still largely different. I'm a neuroscientist; she's an expert cupcake maker. This may seem like I'm being condescending, but I'm really not. Ask anyone what they'd prefer: a discussion on the scientific workings of the brain or a cupcake. See which one is more popular.

The point of this anecdote is to show that two people with very similar origins, environments and genetics can still have vastly different personalities. So what chance does anyone have of predicting and measuring the personalities of two total strangers from the general population?

Take fingerprints. Fingerprints are basically the pattern of ridges in the skin at the end of our digits. Yet, despite this simplicity, almost every human on earth has unique fingerprints. If surface patterns of small patches of skin offer enough variety for everyone to have his or her own exclusive set, how much more variety is possible with something that is the result of countless subtle connections and complex features of the human brain, the most complicated thing in the universe? Even to attempt to determine someone's personality with a simple tool like a written test should be utterly futile, a task akin to sculpting Mount Rushmore with a plastic fork.

However, current theories argue there are predictable and recognisable components of personalities, labelled 'traits', that can be identified via analysis. Just as billions of fingerprints conform to just three types of pattern (loops, whorls and arches) and the vast diversity of human DNA is produced by sequences of just four nucleotides (G, A, T, C), many scientists argue that personalities can be viewed as specific combinations and expressions of certain traits, shared by all people. As J. P. Gillard said in 1959,[2] 'An individual's personality, then, is his unique pattern of traits.' Note how it says 'his'; this was the 1950s, and of course, women were allowed to have personalities only from the mid-1970s.

But what are these traits? How do they combine to form a personality? Arguably the most dominant approach at present is the 'Big 5' personality traits, which argues that there are five traits in particular that make up a personality, in the same way that multiple colours can be made by combining red, blue and yellow. These traits are often consistent across situations and result in predictable attitudes and behaviours in an individual.

Everyone supposedly falls between two extremes of the Big 5 traits:

Openness reflects how open to new experiences you are. If invited to see a new exhibition of sculptures made out of rotten pork, people at the extremes of openness may say, 'Yes, definitely! I've never witnessed art made of rancid meat, so this will be brilliant!' Or, 'No, it's in a different part of town to where I usually am so I won't enjoy it.'

Conscientiousness reflects the extent to which someone is prone to planning, organising, self-discipline. A very conscientious type might agree to attend the rotten-pork exhibition, after working out which would be the best bus route with alternatives in case of traffic disruptions, and also getting a tetanus booster. A non-conscientious type would just agree to meet there in ten minutes, not ask permission to leave work early and opt to follow their nose to find the location.

Extroverts are outgoing, engaging, attention-seeking, while introverts are quiet, private and more solitary. If invited to the rotten-pork exhibition, an extreme extrovert will attend and bring their own hastily made sculpture to show off, and end up posing alongside all the exhibits for their Instagram account. An extreme introvert wouldn't talk to someone long enough to be invited.

Agreeableness reflects the extent to which your behaviour and thinking is affected by a desire for social harmony. A very agreeable person would surely agree to attend the rotten-pork sculpture exhibition, but only as long as the person inviting didn't mind (they don't want to be a bother). Someone totally lacking in agreeableness probably wouldn't be invited anywhere by anyone in the first place.

A neurotic person is invited to a rotten-pork sculpture exhibit and they decline and explain why in exquisite detail. See: Woody Allen.

Unlikely art exhibitions aside, these are the traits that make up the Big 5. There's a lot of evidence to suggest they're quite consistent: a person who scores high on agreeableness will show the same tendencies in a wide variety of situations. There is also some data linking certain personality traits with specific brain activity and regions. Hans J. Eysenck, one of the big names in personality studies, claimed that introverts have higher levels of cortical arousal (stimulation and activity in the cortex) than extroverts.[3] One interpretation of this is that introverts don't require much stimulation. Extroverts, by contrast, want to be excited more often, and develop personalities around this.

Recent scanning studies, like those by Yasuyuki Taki and

others,[4] suggest that individuals demonstrating neuroticism show smaller-than-average areas such as the dorsomedial prefrontal cortex and the left medial temporal lobe including the posterior hippocampus, with a bigger mid-cingulate gyrus. These regions are implicated in decision-making, learning and memory, suggesting a neurotic person is less able to control or suppress paranoid predictions and learn that these predictions are unreliable. Extroversion showed increased activity in the orbitofrontal cortex, which is linked to decision-making, so perhaps because of this raised activity in the decision-making regions, extroverts are compelled to be active and make decisions more often, leading to more outgoing behaviour as a result?

There is also evidence to suggest there are genetic factors underlying personality. A 1996 study by Jang, Livesley and Vernon using nearly 300 pairs of twins (identical and non-identical) suggested that the heritability of the Big 5 personality traits ranged from between 40 per cent to 60 per cent.[5]

What the preceding paragraphs boil down to is that there are some personality traits, specifically five, that have a large body of evidence behind them and appear to be associated with brain regions and genes. So what's the issue?

Firstly, many argue that the Big 5 personality traits don't provide a thorough description of the true complexity of personality. It's a good overall range, but what about humour? Or tendency to religion or superstition? Or temper? Critics suggest the Big 5 are more indicative of 'outward' personality; all those traits can be observed by another person, whereas much of personality is internal (humour, beliefs, prejudices and so on), taking place largely inside your head and not necessarily being reflected in behaviour.

We've seen evidence that personality types are reflected in the configuration of the brain, suggesting they have biological origins. But the brain is flexible and changes in response to what it experiences, so the brain configurations we see could be a consequence of the personality types, not a cause. Being very neurotic or extroverted means you end up with distinct experiences, which could be what the arrangement of your brain bits is reflecting. This is assuming the data itself is 100 per cent confirmed, which it isn't.

There's also the manner of how the Big 5 theory came about. It is based on factor analysis (discussed in Chapter 4) of data produced by decades of personality research. Many different analyses by different people have found these five traits repeatedly, but what does this mean? Factor analysis just looks at the available data. Using factor analysis here is like putting several large buckets across town in order to collect rain. If one persistently fills up before the others, you can say the location of that bucket gets more rain than elsewhere. This is good to know, but it doesn't tell you *why*, or how rain forms, or the various other important aspects. It's useful information, but it's just the start of understanding, not the conclusion.

The Big-5 approach has been focused on here because it's the most widespread, but it's far from the only one. In the 1950s, Friedman and Rosenhan came up with Type-A and Type-B personalities,[6] with Type-As being competitive, achievement-seeking, impatient and aggressive, and Type-Bs not being these things. These personality types were linked to the workplace, as Type-As often end up in management or high-flying positions due to their characteristics, but a study found that Type-As were twice as likely to suffer from heart attacks or other cardiac ailments. Having a personality type could literally kill you, which wasn't encouraging. But follow-up studies suggested this tendency towards heart failure was due to other factors, such as smoking, poor diet, the strain of screaming at subordinates every eight minutes and so on. This Type-A/Type-B approach to personality was found to be too generalised. A more subtle approach was needed, hence the more detailed interest in traits.

Much of the actual data that trait theories emerged from was based on linguistic analysis. Researchers including Sir Francis Galton in the 1800s and Raymond Cattell (the man behind fluid and crystallised intelligence) in the 1950s looked at the English language and assessed it for words that revealed personality traits. Words such as 'nervous', 'anxious' and 'paranoid' can all be used to describe neuroticism, whereas words such as 'sociable', 'friendly' and 'supportive' can apply to agreeableness. Theoretically, there can be only as many terms of this kind as there are personality traits to apply them to – the so-called Lexical Hypothesis.[7] The descriptive words were all collated and crunched and the specific personality types emerged from it, and provided a lot of data for the formation of later theories.

There are problems with this approach too, primarily as it depends on language, something that varies between cultures and is constantly in flux. Other more sceptical types argue that approaches such as the trait theory are too restrictive to be truly representative of a personality: nobody behaves the same way in all contexts; the external situation matters. An extrovert may be outgoing and excitable, but if they're at a funeral or an important business meeting they wouldn't behave in an extroverted manner (unless they've got deep-seated issues), so they would handle each occasion differently. This theory is known as situationism.

Despite all the scientific debate, personality tests are common.

Completing a quick quiz, and then being told you conform to a certain type, is a bit of fun. We feel we have a certain type of personality, and completing a test that says we do have this type validates our assumptions. It might be a free test on some poorly assembled website that keeps asking us to sign up to an online casino every six seconds, but a test is a test. The classic is the Rorschach test, where you look at an unspecified pattern of blobs and say what you see, such as 'butterflies emerging from a cocoon', or 'the exploded head of my therapist who asked me too many questions'. While this might reveal something of an individual's personality, it isn't something that can be verified. A thousand very similar people could look at the same image and give a thousand different answers. Technically, this is a very accurate demonstration of the complexity and variability of personality, but it's not scientifically useful.

But it's not all frivolous. The most worrying and widespread use of personality tests is in the corporate world. You may be familiar with the Myers-Briggs Type Inventory (MBTI), one of the most popular personality-measuring tools in the world, worth millions of dollars. The trouble is, it is not supported or approved by the scientific community. It looks rigorous and sounds proper (it too relies on scales of traits, extrovert–introvert being the most well-known one), but it's based on untested decades-old assumptions put together by enthusiastic amateurs, working from a single source.[8] Nonetheless, at some point it was seized on by business types who wanted to manage employees in the most effective manner, and thus it became globally popular. It now has hundreds of thousands of proponents who swear by it. But then, so do horoscopes.

One explanation for this is the MBTI is relatively straightforward and easily understood, and allows sorting of employees into useful categories that help predict their behaviour and manage them accordingly. You employ an introvert? Put her in a position where she can work alone and don't disturb her. Meanwhile, take the extroverts and put them in charge of publicity and engagement; they like that.

At least, that's the theory. But it can't possibly work in practice, because humans are nowhere near that simple. Many corporations use the MBTI as an integral component of their hiring policies, a system that relies on the applicant being 100 per cent honest and almost as clueless. If you're applying for a job and they make you do a test which asks, 'Do you enjoy working with others?', you're unlikely to put, 'No, others are vermin, only there to be crushed', even if you do think this. The majority of people have sufficient intelligence to play it safe with such tests, thus rendering the results meaningless.

The MBTI is regularly used as an irrefutable gold standard by non-scientific types who don't know better and have been caught up in the hype. The MBTI being infallible could only ever be the case if everyone who completed it actively played along with their personality diagnoses. But they won't. The fact that it would be helpful for managers if people conformed to limited and easily understood categories doesn't mean it's what happens.

Overall, personality tests would be more useful if our personalities didn't get in the way.

Do blow your fuse (How anger works and why it can be a good thing)

Bruce Banner has a famous catchphrase: 'Don't make me angry. You wouldn't like me when I'm angry.' When Banner becomes angry, he turns into the Incredible Hulk, world-famous comic-book character beloved by millions. So the catchphrase is clearly untrue.

Also, who *does* like someone when they're angry? Granted, some people display 'righteous fury' when they get fired up about an injustice, and those who agree will cheer them on. But anger is generally seen as a negative, largely because it produces irrational behaviour, upset and even violence. If it's so harmful, why is the human brain so keen to produce it in response to even the most irrelevant-seeming occurrence?

What exactly is anger? A state of emotional and physiological arousal, typically experienced when some sort of boundary is violated. Someone collides with you in the street? Your physical boundary has been violated. Someone borrows money from you and won't give it back? Your financial or resource boundary has been violated. Someone expresses views you find incredibly offensive? Your moral boundary has been violated. If it is obvious that whoever has violated your boundary has done so on purpose, this is provocation, and results in even greater levels of arousal, thus more anger. It's the difference between spilling someone's drink and actively throwing it in their face. Not only have your boundaries been violated; someone did it deliberately, for their benefit at your expense. The brain has been responding to trolls since long before the Internet.

The recalibration theory of anger, put forward by evolutionary psychologists,[9] argues that anger evolved to deal with scenarios like this, as a sort of self-defence mechanism. Anger provides a quick subconscious way of reacting to a situation that has caused you to lose out, making it more likely you'll address the balance and ensure self-preservation. Imagine a primate ancestor, painstakingly making a stone axe via his newly evolved cortex. It takes time and effort to make these new-fangled 'tools', but they are useful. Then, once completed, someone comes and takes it for himself. A primate that responds by quietly sitting and mulling on the nature of possession and morality may seem the smarter one, but the one that gets angry and punches the thief in the jaw with his ape-like fists gets to keep his tool and is far less likely to be disrespected again, thus increasing his status and chances of mating.

That's the theory, anyway. Evolutionary psychology does seem to have a habit of oversimplifying things like this, which itself angers people.

In a strictly neurological sense, anger is often the response to a threat, and the 'threat-detection system' is strongly implicated in anger. The amygdala, hippocampus and periaqueductal grey, all regions of the midbrain responsible largely for fundamental processing of sensory information, make up our threat-detection system, and thus have roles in triggering anger. However, the human brain, as we saw earlier, keeps using the primitive threat-detection system to navigate the modern world and considers being laughed at by colleagues because a co-worker keeps doing unflattering impressions of you as a 'threat'. This doesn't harm you in any physical sense, but your reputation and social standing are at risk. End result, you get angry.

Brain-scanning studies, such as those conducted by Charles Carver and Eddie Harmon-Jones, have shown that subjects who are angered demonstrate raised levels of activity in the orbitofrontal cortex, a brain region often associated with the control of emotions and goal-orientated behaviour.[10] This basically means that when the brain wants something to happen, it induces or encourages behaviour that will cause this thing to happen, often via emotions. In the case of anger, something happens, your brain experiences it, decides that it's really not happy about it, and produces an emotion (anger) in order to respond and effectively deal with it in a satisfactory manner.

Here's where it gets more interesting. Anger is seen as destructive and irrational, negative and harmful. But it turns out anger is sometimes useful, indeed helpful. Anxiety and threats (of many sorts) cause stress, which is a big problem, largely because it triggers release of the hormone cortisol, producing the unpleasant physiological consequences that make stress so harmful. But many studies, such as that done by Miguel Kazén and his colleagues for Universität Osnabrück,[11] show that experiencing anger *lowers* cortisol, thus reducing the potential harm caused by stress.

One explanation for this is that studies* have shown anger causes raised activity in the left hemisphere of the brain, in the anterior cingulate cortex in the middle of the brain, and

* As an aside, it's worth noting that studies into anger report doing things like 'presenting subjects with stimuli designed to raise levels of anger', but a lot of the time this means they're basically just insulting the volunteers. It's understandable why they'd not want to reveal this too openly; psychological experiments invariably rely on people volunteering to take part, and they're less likely to do that if they find it involves being strapped to a scanner while a scientist uses colourful metaphors to tell you how fat your mother is.

the frontal cortex. These regions are associated with producing motivation and responsive behaviour. They are present in both brain hemispheres, but do different things on each side; in the right hemisphere they produce negative, avoidance or withdrawal reactions to unpleasant things, and in the left hemisphere produce positive, active, approach behaviour.

To put it simply, when it comes to this motivational system being presented with a threat or a problem, the right half says, 'No, stay back, it's dangerous, don't make it worse!', causing you to recoil or hide. The left half says, 'No, I'm not having this, it needs to be dealt with', before metaphorically rolling up its sleeves and getting stuck in. The metaphorical devil and angel on your shoulder are actually lodged in your head.

People with a more confident, extroverted personality probably have a dominant left side, while for neurotic or introverted types it's likely to be the right. But the right side's influence doesn't lead to anything being done about apparent threats, so they persist, causing anxiety and stress. Available data suggests that anger increases activity in left hemisphere system,[12] potentially prompting someone into action in the manner of someone shoving a hesitant person off a diving board. Lowering cortisol at the same time limits the anxiety response that can 'freeze' people. Eventually dealing with the stress-causing thing lowers cortisol further.* Similarly, anger has also been shown to make people think more optimistically, so rather than fearing the worst

* The same studies demonstrated that anger hinders performance on complex cognitive tasks, showing how anger means you can't 'think straight'. Not always helpful, but it would inevitably feed into the same system. You could calmly assess all the properties of the threat you encounter and decide that, overall, it's too risky to deal with. But anger hinders this rational thinking, messing up the delicate analysis that leads to you avoid the issue and compels you to go right at it, fists flailing.

from a potential outcome, it encourages people to think any issue can be dealt with (even if that's wrong), so any threat is minimised.

Studies have also shown that visible anger is useful in negotiations, even if both parties are showing it, as there's more motivation to obtain something, greater optimism as to the outcome, and an implied honesty to all that is said.[13]

All this disputes the idea that you should bottle up anger, and suggests you should instead let it out in order to reduce stress and get things done.

But, as ever, anger is not so simple. It comes from the brain, after all. We've developed many ways to suppress the anger response. The classic 'count to ten' or 'take deep breaths before responding' strategies make sense when you consider the anger response is very quick and intense.

The orbitofrontal cortex, highly active during experiences of anger, is involved with the control of emotions and behaviour. More specifically, it modulates and filters emotional influence over behaviour, damping down or blocking our more intense and/or primitive impulses. When an intense emotion is most likely to cause us to behave dangerously, the orbitofrontal cortex steps in as a sort of stopgap, acting like the overflow outlet on a bathtub with a leaky tap; it doesn't address the underlying problem, but stops it from getting too bad.

The immediate visceral sensation of anger isn't always the extent of it. Something that angers you can leave you seething for hours or days, even weeks. The initial threat-detection system leading to anger involves the hippocampus and amygdala, areas we know are involved in forming vivid and emotionally charged memories, so the anger-causing occurrence would persist in the memory, leading us to dwell on it, or 'ruminate' to give it the official term. Subjects ruminating on something that made them angry show increased activity in the medial prefrontal cortex, another area involved in making decisions, plans and other complex mental actions.

As a result, we often see anger persisting, even building up. This is especially the case for minor irritations we have no response for. Anger may make your brain want to address the aggravating problem, but what if it's a vending machine that didn't give you any change? Or someone recklessly cut you up on the motorway? Or your boss saying you need to work late at 4.56 p.m.? All of these cause anger but there are no options for dealing with them, unless you want to commit vandalism/crash your car/get fired. And these things can all happen on the same day. So now your brain is in a state of having multiple angering things to dwell on and no obvious options to deal with them. The left-hand element of your behavioural response system is urging you to do something, but what is there to do?

Then a waiter accidentally brings you a black coffee instead of a latte and then that's your limit. The hapless service person gets both barrels of an enraged tirade. This is 'displacement'. The brain has all this anger built up but no outlet, and transfers it onto the first viable target it encounters, just to release the cognitive pressure. This doesn't make it any more pleasant for the person who unintentionally opened the furious floodgates.

If you are angry and don't want to show it, the brain's versatility means there are ways to be aggressive without using crude violence. You can be 'passive aggressive', where you make another person's life miserable via behaviour they can't really object to. Talking to them less or speaking to them neutrally when you're normally quite friendly, inviting all your mutual friends to social events but not them; neither of these behaviours are definitely hostile, but as a result they lead to uncertainty. The other person is upset or uncomfortable but they can't say for sure if you're angry at them, and the human brain doesn't like ambiguity or uncertainty; it finds them distressing. Thus the other person is punished without violence or violation of social norms.

This passive-aggressive method can work because humans are very good at recognising when another person is angry. Body language, expression, tone of voice, chasing you with a rusty machete while screaming; your typical brain can pick up on all these subtle cues and deduce anger. This can be helpful, as people don't like it when others are angry; it means they present a possible threat or may behave in harmful or upsetting ways. But it also reveals that something has genuinely aggrieved that person.

Another important thing to remember is the experience of anger and the response to anger are not the same thing. The sensation of anger is arguably the same for everyone, but how people react to it varies substantially, another indication of personality type. The emotional response when someone threatens you is anger. Should you respond by behaving in a manner that will harm whoever's responsible, this is *aggression*. To round it off, the thinking about causing harm to someone is *hostility*, the cognitive component of aggression. You catch a neighbour painting a swear word onto your car, you experience anger. You think, 'I'm going to absolutely batter them for this' – that's hostility. You throw a brick through their front window in response, that's aggression.*

* Aggression can also happen without anger. Contact sports such as rugby or football often involve aggression, but no anger is required; it's just the desire to win at the expense of the other team that motivates it.

So should we let ourselves get angry or not? I'm not suggesting you go and row with colleagues or force them through the office shredder every time they irritate you, but be aware that anger isn't always a bad thing. However, moderation is key. Angry people tend to have their needs addressed before people who make polite requests. This means you get people who realise that being angry benefits them, so they do it more often. The brain eventually associates constant anger with rewards, so encourages it further, and you end up with someone who gets angry at the slightest inconvenience just to get their own way, and then they inevitably become a celebrity chef. Whether that's a good or bad thing is down to you.

Believe in yourself, and you can do anything . . . within reason

(How different people find and use motivation)

'The harder the journey, the better the arrival.'

'Effort is just the foundation of a house that is you.'

These days you can't enter a gym or coffee shop or workplace canteen without being exposed to several insipid motivational posters featuring quotes like this. The previous section on anger discussed how that emotion can motivate someone to respond to a threat in a specific way via dedicated brain pathways, but we're talking here about more long-term motivation, the kind that's more a 'drive' than a reaction.

What is motivation? We know when we aren't motivated – many assignments have been scuppered by the person responsible procrastinating. Procrastination is motivation to do the wrong thing (I should know, I had to disconnect my wifi to finish this book). Broadly, motivation can be described as the 'energy' required for a person to remain interested and/or working towards a project, goal or outcome. An early theory of motivation comes from Sigmund Freud himself. Freud's hedonic principle, sometimes called the 'pleasure principle', argues that living beings are compelled to seek out and pursue things that give pleasure, and avoid things that cause pain and discomfort.[14] That this happens is hard to deny, as studies into animal learning have shown. Put a rat in a box and give it a button, it'll press it eventually out of sheer curiosity. If pressing the button results in a tasty food being supplied, the rat will quickly start pressing the button often because it's associated doing this with a tasty reward. It's not a stretch to say it's suddenly very motivated to press the lever.

This very reliable process is known as operant condition, meaning a certain type of reward increases or decreases the specific behaviour associated with it. This occurs in humans too. If a child is given a new toy when they clean their room, they're far more likely to want to do it again. It also works with adults, too; you just need to vary the reward. As a result, the unpleasant task of cleaning a room is now associated with a positive outcome, so there's motivation to do it.

This may all seem to support Freud's hedonic principle, but when have humans and their irksome brains ever been so simple? There are plenty of everyday examples to demonstrate there's more to motivation than simple pleasure-seeking or displeasure-avoiding. People are constantly doing things that provide no immediate or obvious physical pleasure.

Take going to the gym. While it is true that intense physical activity can produce euphoria or feelings of well-being,* this doesn't happen every time, and it still takes gruelling effort to get to that point, so there's no obvious physical pleasure to be had from exercise (I say this as someone who's yet to experience so much as a satisfying sneeze from going to the gym). And yet, people still do it. Whatever their motivation, it is clearly something beyond immediate physical pleasure.

There are other examples. People who regularly give to charity, surrendering their own money for the benefit of strangers they'll never encounter. People who constantly suck up to a deeply unpleasant boss in the vague hope of getting a promotion. People reading books they don't really enjoy but persevering regardless because they want to learn something. None of these things involve immediate pleasure; some actually involve unpleasant experiences, so according to Freud they would be avoided. But they aren't.

This suggests Freud's ideas are too simplistic,** so a more

* Exactly why this 'runner's high' occurs is uncertain. Some say it's using up the muscle's oxygen supplies, triggering anaerobic respiration (oxygen-free cellular activity, which produces acid by-products that can cause pain, such as cramps or a 'stitch'), which the brain responds to by releasing endorphins, the pain-killing pleasure-inducing transmitters. Others say it's more to do with elevated body temperature, or constant rhythmic activity providing a sense of well-being that the brain wants to encourage. Marathon runners often report this runner's high, which as a rewarding sensation seems to come second only to the telling people, 'I'm training for a marathon you know', given how often they find excuses to do this.

** Freud still has a lot of influence and many adhere to his theories, even a century later. This may seem odd. Granted, he did largely usher in the whole concept of psychoanalysis and should be lauded for it, but this doesn't mean his original theories are automatically correct. It is the diffuse and uncertain nature of psychology and psychiatry that means he still wields such influence today; it's hard to disprove things conclusively. Yes, Freud founded the whole field, but the Wright brothers invented aeroplanes, and while they'll always be remembered for this, we don't still use aircraft that they designed for long-haul flights to South America. Times move on, and all that.

complex approach is needed. You could substitute 'immediate pleasure' with 'needs'. In 1943, Abraham Maslow devised his 'hierarchy of needs', arguing that there were certain things that all humans needed in order to function, and so are motivated to obtain them.[15]

Maslow's hierarchy is often presented as a stepped pyramid. At the lowest level are biological requirements such as food, drink, air (someone without air is undeniably very motivated to find some). Then there's safety, including shelter, personal security, financial security, things that stop you from coming to physical harm. Next is 'belonging'; humans are social creatures and need approval, support and affection (or at least interaction) from others. Solitary confinement in prisons is considered a serious punishment for a reason.

Then there's 'esteem', the need to be not just acknowledged or liked but actually respected by others, and by yourself. People have morals that they value and stick to, and hope others will respect them for. Behaviour and actions that can lead to this are therefore a source of motivation. Finally, there's 'self-actualisation', the desire (and therefore motivation) to reach one's potential. You feel you could be the best painter in the world? Then you will be motivated to become the best painter in the world. Although, since art is subjective, you technically may already be the best painter in the world. Well done, if so.

The idea is that a person would be motivated to meet all the needs of the first level, then the second level, then the third and so on, in order to satisfy all needs and drives and be the best possible person. It's a nice idea, but the brain isn't that neat and organised. Many people don't follow Maslow's hierarchy; some are motivated to give the last of their money to help strangers in need, or actively put themselves in harm's way to save an animal in danger (unless it's a wasp), despite the fact that an animal has no means of respecting or rewarding them for their heroics (especially if it's a wasp, which will probably sting them and do an evil wasp laugh).

There's also sex. Sex is a very powerful motivator. For proof of this, see anything ever. Maslow states that sex is at the bottom of the hierarchy of needs, as it's a primitive, powerful biological drive. But people can live without any sex at all. They might resent doing so, but it's entirely possible. Also, why do people want sex? A primitive urge for pleasure and/or reproduction, or the desire to be close and intimate with someone? Or maybe it's because others view sexual prowess as an achievement and deserving of respect? Sex is all over the hierarchy.

Recent research into the workings of the brain provide another approach to understanding motivation. Many scientists draw distinctions between intrinsic and extrinsic motivation. Are we being motivated by external factors, or internal ones? External motivations are derived from others. Someone pays you to help them move house; that's an external motivation. You won't enjoy it, it's tedious and involves heavy lifting, but you get rewarded financially and so you do it. It could also be more subtle. Say everyone starts wearing yellow cowboy hats for 'fashion', and you want to be trendy, so you buy and wear a yellow cowboy hat. You may have no fondness for yellow cowboy hats, you may think they look stupid, but others have decided otherwise, and so you want one. This is an extrinsic motivation.

Intrinsic motivations are where we're driven to do things because of decisions or desires that we come up with ourselves. We decide, based on what we've experienced and learned, that helping sick people is a noble and rewarding thing to do, so we're motivated to study medicine and become a doctor. This is an intrinsic motivation. If we are motivated to study medicine because people pay doctors a lot of money, this is more an extrinsic motivation.

Intrinsic and extrinsic motivations exist in a delicate balance. Not with each other, but within themselves as well. In 1988, Deci and Ryan came up with the self-determination theory, which describes what motivates people in the absence of any external influence, so is 100 per cent intrinsic.[16] It argues that people are motivated to achieve autonomy (control of things), competency (to be good at things) and relatedness (be recognised for what they do). All of these explain why micromanagers are so infuriating; someone hovering over your shoulder telling you precisely how to do the simplest task robs you of all control, undermines all notion of competence and is often impossible to relate to, given how most micromanagers seem sociopathic (if you're at the mercy of one).

In 1973, Lepper, Greene and Nisbet pointed out the over-justification effect.[17] Groups of children were given colourful art supplies to play with. Some of were told they'd be rewarded for using of them; others were left to their own devices. A week later, the children who *weren't* rewarded were far more motivated to use the art supplies again. Those who decided that the creative activity was enjoyable and satisfying on their own experienced greater motivation than those who received rewards from other people.

It seems if we associate a positive outcome with our own actions, this carries more weight than if the positive outcome came from someone else. Who's to say they won't reward us next time? As a result, motivation is diminished.

The obvious conclusion is that rewarding people for a task can actually *reduce* motivation for doing it, whereas giving them more control or authority increases motivation. This idea has been picked up (with great enthusiasm) by the business world, largely because it lends scientific credibility to the idea that it's better to give employees greater autonomy and responsibility than actually paying them for their labour. While some researchers suggest that this is accurate, there's ample data against it. If paying someone to work reduces motivation, then top executives who get paid millions actually do nothing. Nobody is saying that though; even if billionaires aren't motivated to do anything, they can afford lawyers who are.

The brain's tendency towards ego can also be a factor. In 1987, Edward Tory Higgins devised the self-discrepancy theory.[18] This argued that the brain has a number of 'selves'. There's the 'ideal' self, which is what you *want* to be, derived from your goals, biases and priorities. You may be a stocky computer programmer from Inverness, but your ideal self is a bronzed volleyball player living on a Caribbean island. This is your ultimate goal, the person you want to be.

Then there's the 'ought' self, which is how you feel you *should* be behaving in order to achieve the *ideal* self. Your 'ought' self is someone who avoids fatty foods and wasting money, learns volleyball and keeps an eye on Barbados property prices. Both selves provide motivation; the ideal self provides a positive kind of motivation, encouraging us to do things that bring us closer to our ideal. 'Ought' self provides more negative, avoidance motivation, to keep us from doing things that take us away from our ideal; you want to order pizza for dinner? That's not what you *ought* to do. Back to the salads for you.

Personality also plays a part. When it comes to motivation,

someone's locus of control can be crucial. This is the extent to which someone feels they are in control of events. They might be an egotistical sort who feels the very planet revolves around them, because why wouldn't it? Or they may be far more passive, feeling they're always at the mercy of circumstance. Such things may be cultural; people raised in a Western capitalist society, constantly told they can have anything they want, will feel more in control of their own lives, whereas someone living in a totalitarian regime probably won't.

Feeling like a passive victim of events can be damaging; it can reduce the brain to a state of learned helplessness. People don't feel they can change their situation, so lack the motivation to try. They don't attempt to do anything as a result, and things get worse for them due to their inaction. This lowers their optimism and motivation further, so the cycle continues and they end up an ineffectual mess, paralysed by pessimism and zero motivation. Anyone who's ever been through a bad break-up can probably relate to this.

Exactly where motivation originates in the brain is unclear. The reward pathway in the midbrain is implicated, along with the amygdala due to the emotional component involved in things that motivate us. Connections to the frontal cortex and other executive areas are also associated as a lot of motivation is based on planning and anticipation of reward. Some even argue that there are two separate motivation systems, the advanced cognitive kind that gives us life goals and ambitions, and the more basic reactive kind that says, 'Scary thing, run!' Or, 'Look! Cake! Eat it!'

But the brain also has other quirks that produce motivation. In the 1920s, Russian psychologist Bluma Zeigarnik noticed, while sitting in a restaurant, that the waiting staff seemed to be able to remember only the orders they were in the process of dealing with.[19] Once the order was completed, they seemed to lose all memory of it. This occurrence was later tested in the lab. Subjects were given simple tasks to do, but some were interrupted before they could complete them. Later assessment revealed that those who were interrupted could remember the tests much better, and even wanted to complete them despite the test being over and there being no reward for doing so.

This all gave rise to what is now known as the Zeigarnik effect, where the brain really doesn't like things being incomplete. This explains why TV shows use cliff-hangers so often; the unresolved storyline compels people to tune into the conclusion, just to end the uncertainty.

It seems as if the second best way to motivate a person to do something is to leave it incomplete and restrict their options for resolving it. There is an even more effective way to motivate people, but that will be revealed in my next book.

Is this meant to be funny?
(The weird and unpredictable workings of humour)

'Explaining a joke is like dissecting a frog. You understand it better but the frog dies in the process' – E. L. White. Unfortunately, science is largely about rigorous analysis and explaining things, so this may be why science and humour are often seen as mutually exclusive. Despite this, scientific attempts have been made to investigate the brain's role in humour. Numerous psychological experiments have been detailed throughout this book: IQ tests, word-recitation tests, elaborate food preparations for appetite/taste, and so on. One of the common properties of these experiments, and countless others used in psychology, is that they all adhere to certain types of manipulations, or 'variables' to use the technical term.

Psychology experiments incorporate two types of variables: independent and dependent variables. Independent variables are what the experimenter controls (IQ test for intelligence, word lists for memory analysis, all designed and/or supplied by the researcher); dependent variables are what the experimenter measures, based on how the subjects respond (score on IQ test, number of things remembered, bits of brain that light up and so on).

Independent variables need to be reliable in invoking the desired response, for example, the completion of a test. And here's where a problem arises; in order to study effectively how humour works in the brain, your subjects need to experience humour. So ideally, you'd need something that *everyone, no matter who they are, is guaranteed to find funny*. Anyone who can come up with such a thing probably won't be a scientist for very long, as they'd soon be getting vast sums from television companies desperate to exploit this skill. Professional comedians work for years to achieve this, but there's never been a comedian that *everyone* likes.

It gets worse, because surprise is a big element of comedy and humour. People will laugh when they first hear a joke they like, but not so much the second, third, fourth or more times they hear it, because now they know it. So any attempt to repeat the experiment* will need yet another 100 per cent reliable way to make people laugh.

* It may seem wasteful or lazy, but repetition is a very important process in science because repeating an experiment and getting the same results helps make

There's also the setting to consider. Most laboratories are very sterile, regulated environments, designed to minimise risks and prevent anything from interfering with experiments. This is great for science, but not for encouraging a state of merriment. And if you're scanning the brain, it's even harder; MRI scans, for example, involve being confined in a tight chilly tube while a massive magnet makes very weird noises all around you. This isn't the best way to put someone in the mood for knock-knock jokes.

But still, a number of scientists haven't let these fairly considerable obstacles stop them investigating the workings of humour, although they've had to adopt some odd strategies. Take Professor Sam Shuster, who looked into the workings of humour and how it differs between groups of people.[20] He did this by riding a unicycle in busy public areas of Newcastle and recording the types of reactions this provoked. While an innovative form of research, on a list of potential candidates for things everybody finds amusing, 'unicycles' is unlikely to be in the top ten.

There's also a study by Professor Nancy Bell of Washington State University,[21] whereby a deliberately bad joke was regularly dropped into casual conversations, in order to determine the nature of people's reactions to poor attempts at humour. The joke used was: 'What did the big chimney say to the little chimney? Nothing. Chimneys can't talk.'

The responses ranged from awkward to outright hostile. Overall though, it seems nobody actually *liked* the joke, so

---

sure that the findings are reliable, not just due to a luck or sneaky manipulation. This is a particularly big problem in psychology, given the unpredictability and unreliability of the human brain. It even thwarts attempts to study it, which is another annoying property of it.

whether this even counts as a study into humour is debatable.

These tests technically look at humour indirectly, via reactions and behaviour towards people attempting it. *Why* do we find things funny? What's going on in the brain to make us respond to certain occurrences with involuntary laughter? Scientists to philosophers have chewed this over. Nietzsche argued that laughter is a reaction to the sense of existential loneliness and mortality that humans feel, although judging by much of his output Nietzsche wasn't that familiar with laughter. Sigmund Freud theorised that laughter is caused by the release of 'psychic energy', or tension. This approach has developed and been labelled the 'relief' theory of humour.[22] The underlying argument is that the brain senses some form of danger or risk (to ourselves or others), and once it is resolved harmlessly, laughter occurs to release the pent-up tension and reinforce the positive outcome. The 'danger' can be physical in nature, or something inexplicable or unpredictable like the twisted logic of a joke scenario, or suppression of responses or desires due to social constraints (offensive or taboo jokes often get a potent laugh, possibly because of this). This theory seems particularly relevant when applied to slapstick; someone slipping on a banana skin and ending up dazed is humorous, whereas someone slipping on a banana skin, cracking their skull and dying is certainly not, because the danger is 'real'.

A theory by D. Hayworth in the 1920s builds on this,[23] arguing that the actual physical process of laughter evolved as a way for humans to let each other know that the danger has passed and all is well. Where this leaves people who claim 'to laugh in the face of danger' is anyone's guess.

Philosophers as far back as Plato suggested that laughter is an expression of superiority. When someone falls over, or does or says something stupid, this pleases us because they have lowered their status compared with ours. We laugh because we enjoy the feeling of superiority and to emphasise the other person's failings. This would certainly explain the enjoyment of *Schadenfreude*, but when you see internationally famous comedians strutting about on stage performing to thousands of laughing people in stadiums, it's unlikely the entire audience is thinking, 'That person is stupid. I am better than them!' So again, this isn't the whole story.

Most theories concerning humour highlight the role of inconsistency and disrupted expectations. The brain is constantly trying to keep track of what's going on both externally and internally, in the world around us and inside our heads. To facilitate this, it has a number of systems to make things easier, such as schemas. Schemas are specific ways that our brains think and organise information. Particular schemas are often applied to specific contexts – in a restaurant, at the beach, in a job interview, or when interacting with certain individuals/types of people. We expect these situations to pan out in certain ways and for a limited range of things to transpire. We also have detailed memories and experiences that suggest how things are 'meant' to occur in recognisable circumstances and scenarios.

The theory is that humour results when our expectations are violated. A verbal joke uses twisted logic, where events don't occur as we believe they should. Nobody has ever gone to the doctor because they feel like a pair of curtains. Unattended horses seldom walk into bars. Humour potentially comes from being faced with these logical or contextual inconsistencies as they cause uncertainty. The brain isn't good at uncertainty, especially if it means the systems it uses to construct and predict our world view are potentially flawed (the brain expects something to happen in a certain way, but it doesn't, which suggests underlying issues with its crucial predictive or analytical functions). Then the inconsistency is resolved or defused by the 'punchline', or equivalent. Why the long face? A horse has a long face, but that's a question asked to miserable people! It's wordplay! I understand wordplay! The resolution is a positive sensation for the brain as the inconsistency is neutralised, and maybe something is learned. We signal our approval of resolution via laughter, which also has numerous social benefits.

This also helps explain why surprise is so important, and why a joke is never as funny when repeated; the inconsistency that caused the humour originally is no longer unfamiliar, so the impact is dulled. The brain now remembers this set-up, is aware that it is harmless, so isn't as affected by it.

Many brain regions have been implicated in the processing of humour, such as the mesolimbic reward pathway, given that it produces the reward of laughter. The hippocampus and amygdala are involved, as we need to have memories of what *should* happen to have these anticipations thwarted, and strong emotional responses to this occurring. Numerous frontal cortex regions play a role, as much of humour comes from expectations and logic being disrupted, which engage our higher executive functions. There are also parietal lobe regions involved in language processing, as much comedy is drawn from wordplay or violating the norms of speech and delivery.

This language-processing role of humour and comedy is more integral than many may think. Delivery, tone, emphasis, timing, all of these can make or break a joke. A particularly interesting finding concerns the laughing habits of deaf people who communicate via sign language. In a standard vocal conversation where someone tells a joke or a humorous story, people laugh (if it's funny) during the pauses, at the ends of sentences, basically in the gaps where laughing will not obscure the telling of the joke. This is important because laughter and joke-telling are usually both sound-based. This isn't the same for sign-language speakers. Someone could laugh throughout a joke or story told via sign language and not obscure anything. But they don't. Studies show that deaf people laugh at the same pauses and gaps during a signed joke, even if the noise of laughter isn't a factor.[24] Language and speech processing clearly influence when we feel it's time to laugh, so it's not necessarily as spontaneous as we think.

As far as we currently know, there is no specific 'laughter centre' in the brain; our sense of humour seems to arise from myriad connections and processes that are the result of our development, personal preferences and numerous experiences. This would explain why everyone has his or her own seemingly unique sense of humour.

Despite the apparent individuality of a person's tastes in comedy and humour, we can prove that it is heavily influenced by the presence and reactions of others. That laughing has a strong social function is undeniable; humans can experience many emotions as suddenly and intensely as humour, but the majority of these emotions don't result in loud uncontrolled (often incapacitating) spasms (i.e. laughter). There is benefit to making your amused state public knowledge, because people have evolved to do this whether they want to or not.

Studies such as those by Robert Provine of the University of Maryland suggest that you are thirty times as likely to laugh when you're part of a group as when you're alone.[25] People laugh more often and freely when with friends, even if they're not telling jokes; it can be observations, shared memories, or very mundane-sounding anecdotes about a mutual acquaintance. It's a lot easier to laugh when part of a group, which is why stand-up comedy is rarely a one-to-one practice. Another interesting point about the social-interaction qualities of humour: the human brain appears to be very good at distinguishing between real laughter and fake laughter. Research by Sophie Scott has revealed people to be extremely accurate when it comes to identifying someone laughing genuinely and someone pretending, even if they sound very similar.[26] Have you ever been inexplicably annoyed by obvious canned laughter on a cheesy sitcom? People respond to laughter very strongly, and they invariably object to this response being manipulated.

When an attempt to make you laugh fails, it fails *hard*.

When someone tells you a joke, they are making it clear that they are intending to make you laugh. They have concluded that they know your humour and are able to make you laugh, and are thereby asserting that they are able to control you, so are superior to you. If they're doing this in front of people, then they're really emphasising their superiority. So it had better be worth it.

But then it's not. The joke falls flat. This is basically a betrayal, one that offends on several (largely subconscious) levels. It's no wonder people often get angry (for examples of this, just ask any aspiring comedian, anywhere, ever). But to appreciate this fully, you have to appreciate the extent to which interactions with others influence the workings of our brains. And that needs a chapter of its own to do it justice.

Only then can it really be grasped, as the actress said to the bishop.

7

Group hug!

How the brain is influenced by other people

Many claim to not care what anyone thinks of them. They will say this often, and loudly, going to great lengths to behave in ways that make it absolutely clear to anyone who'll listen. Apparently, not caring what people think of you isn't valid unless people, the ones you supposedly don't care about, know about it. Those who shun 'social norms' invariably just end up as part of a different recognisable group. From the mods and skinheads of the mid-twentieth century to goths and emos today, the first thing someone does when they don't want to conform to normal standards is to find another group identity to conform to instead. Even biker gangs or the Mafia all tend to dress alike; they may have no respect for the law, but they want the respect of their peers.

If hardened criminals and outlaws can't fight the urge to form groups, it must be quite deeply rooted in our brains. Placing a prisoner in solitary confinement for too long is considered a form of psychological torture,[1] demonstrating that human contact is more a necessity than a desire. The truth is, odd as it may seem, much of the human brain is dedicated to and formed by interactions with other people, and we grow to depend on people as a result, to a surprising extent.

There's the classic argument about what makes a person

who they are – nature or nurture? Genes or environment? It's a combination of both; genes obviously have a big impact on how we end up, but so do all the things that happen to us as we develop and, for the developing brain, one of if not *the* main source of information and experience is other humans. What people tell us, how they behave, what they do and think/suggest/create/believe, all of this has a direct impact on a still-forming brain. On top of this, much of our selves (self-worth, ego, motivation, ambition and so on) is derived from how others think and behave towards us.

When you consider that other people influence our brain's development, and they are in turn being controlled by their brains, there's only one possible conclusion: *human brains are controlling their own development!* Much apocalyptic sci-fi is based on the idea of computers doing exactly this, but it's not as scary if it's brains doing it because, as we've seen repeatedly, human brains are quite ridiculous. As a result, so are people. And thus we have large portions of our brains dedicated to engaging with others.

What follows are numerous examples of how bizarre this arrangement can end up being.

Written all over your face
(Why it's hard to hide what you're really thinking)

People don't like it when you have a miserable facial expression, even if there's good reason for it, like having had a big row with your partner, or realising you've stepped in dog mess. But, whatever the cause, it's often made worse by some random stranger telling you to smile.

Facial expressions mean other people can tell what someone is thinking or feeling. It's mind reading, but via the face. It's actually a useful form of communication, which shouldn't come as a shock as the brain has a surprisingly extensive variety of processes dedicated to communicating with others.

You may have heard the claim that '90 per cent of communication is non-verbal'. The '90 per cent' claim varies considerably depending on who's saying it, but in truth it varies because people communicate differently in different contexts; people trying to communicate in a crowded nightclub use different methods from those they'd opt for when trying to communicate while trapped in a cage with a sleeping tiger. The overall point is that much or most of our interpersonal communication is conducted via means other than spoken words.

We have several brain regions dedicated to language processing and speech, so the importance of verbal communication should go without saying (ironically). For many years, it was all attributed to two brain regions. Broca's area, named for Pierre Paul Broca, at the rear of the frontal lobe, was believed to be integral to speech formation. Thinking of something to say and putting the relevant words in the correct order, that was Broca's area at work.

The other region was Wernicke's area, identified by Carl Wernicke, in the temporal lobe region. This was credited with language comprehension. When we understand words, their meanings and numerous interpretations, this was the doing of Wernicke's area. This two-component set-up is a surprisingly straightforward arrangement for the brain, and indeed the language system of the brain is actually considerably more complex. But, for decades, Broca's and Wernicke's areas were credited with speech processing.

To understand why, consider that these areas were identified in the nineteenth century, via studies of people who had suffered damage localised to these brain regions. Without modern technology such as scanners and computers, aspiring neuroscientists were reduced to studying unfortunate individuals with just the right sort of head injury. Not the most efficient method, but at least they weren't inflicting these injuries on people themselves (as far as we know).

Broca's and Wernicke's areas were identified because damage to them caused aphasias, which are profound disruptions to speech and understanding. Broca's aphasia, aka expressive aphasia, means someone cannot 'produce' language. There's nothing wrong with their mouth or tongue, they can still understand speech, they just can't produce any fluid, coherent communication of their own. They may be able to utter a few relevant words, but long complex sentences are practically impossible.

Interestingly, this aphasia is often evident when speaking, *or writing*. This is important. Speech is aural and conveyed via the mouth; writing is visual and uses hands and fingers, but for both to be equally impaired means a common element is disrupted, which can be only the language processing, which must be handled separately by the brain.

Wernicke's aphasia is essentially the opposite problem. Those afflicted don't seem able to comprehend language. They can apparently recognise tone, inflection, timing and so on but the words themselves are meaningless. And they respond similarly, with long, complex-sounding sentences, but instead of 'I went to the shops, bought some bread', it's

'I wendle to the do the shops hops todayhayhay boughtage soughtage some read bread breed'; a combination of real and made-up words strung together with no recognisable linguistic meaning, because the brain is damaged in such a way that it cannot recognise language, so also can't produce it.

This aphasia also often applies to written language, and the sufferers are generally unable to recognise any problem with their speech. They think they are speaking normally, which obviously leads to serious frustration.

These aphasias led to the theories about the importance of Broca's and Wernicke's areas for language and speech. However, brain-scanning technology has changed matters. Broca's area, a frontal lobe region, is still important for processing syntax and other crucial structural details, which makes sense; manipulating complex information in real-time describes much frontal lobe activity. Wernicke's area, however, has been effectively demoted due to data that shows the involvement of much wider areas of the temporal lobe around it in processing speech.[2]

Areas such as the superior temporal gyrus, inferior frontal gyrus, middle temporal gyrus and 'deeper' areas of the brain including the putamen are all strongly implicated in speech processing, handling elements such as syntax, the semantic meaning of words, associated terms in memory, and so on. Many of these are near the auditory cortex, which processes how things sound, which makes sense (for once). Wernicke's and Broca's areas may not be as integral for language as first assumed, but they're still involved. Damage to them still disrupts the many connections between language-processing regions, hence aphasias. But that language-processing centres are so widely spread throughout shows language to be a fundamental function of the brain, rather than something we pick up from our surroundings.

Some argue that language is even more neurologically important. The theory of linguistic relativity claims that the language a person speaks underlies their cognitive processing and ability to perceive the world.[3] For instance, if people were raised to speak a language that had no words for 'reliable', then they would be unable to understand or demonstrate reliability, and thus be forced to find work as an estate agent.

This is an obviously extreme example, and it's hard to study because you'd need to find a culture that uses a language with some important concepts missing. (There have been numerous studies into more isolated cultures that have smaller ranges of labels for colours that argue they are less able to *perceive* familiar colours, but these are debatable.[4]). Still, there are many theories about linguistic relativity, the most famous of which is the Sapir–Whorf hypothesis.*

Some go further, claiming that changing the language someone uses *can change how they think*. The most prominent example of this is neuro-linguistic programming, NLP. NLP is a mishmash of psychotherapy, personal development and other behavioural approaches, and the basic premise is that language, behaviour and neurological processes are all intertwined. By altering someone's specific use and experience of language their thinking and behaviour can be changed (hopefully for the better), like someone editing the code for a

* The Sapir–Whorf hypothesis is something of an annoyance to linguists, because it is a very misleading label. The supposed originators, Edward Sapir and Benjamin Lee Whorf, never actually co-authored anything, and never put forward a specific hypothesis. In essence, the Sapir–Whorf hypothesis didn't exist until the term itself was coined, making it a very good example of itself. Nobody said linguistics had to be easy.

computer program to remove bugs and glitches.

Despite its popularity and appeal, there's little evidence to suggest that NLP actually works, putting it in the realms of pseudoscience and alternative medicine. This book is filled with examples of how the human brain does its own thing despite everything the modern world can throw at it, so it's hardly going to fall in line when faced with a carefully chosen turn of phrase.

However, NLP does often state that the non-verbal component of communication is very important, which is true. And non-verbal communication manifests in many different ways.

In Oliver Sacks's seminal 1985 book *The Man Who Mistook His Wife for a Hat*,[5] he describes a group of aphasia patients who cannot understand spoken language, who are watching a speech by the president and finding it hilarious, which is clearly not the intent. The explanation is that the patients, robbed of their understanding of words, have become adept at recognising non-verbal cues and signs that most people overlook, being distracted by the actual words. The president, to them, is constantly revealing that he is being dishonest via facial tics, body language, rhythm of speech, elaborate gestures and so on. These things, to an aphasia patient, are big red flags of dishonesty. When coming from the most powerful man in the world, it's either laugh or cry.

That such information can be gleaned non-verbally isn't a surprise. As previously stated, the human face is an excellent communication device. Facial expressions are important: it's easy to tell when someone is angry, happy, fearful and so on because their face takes on an associated expression revealing this, and this contributes greatly to interpersonal communication. Someone could say, 'You shouldn't have', while looking happy, angry or disgusted, and the phrase would be interpreted very differently.

Facial expressions are quite universal. Studies have been conducted where pictures of specific facial expressions have been shown to individuals from different cultures, some of which were very remote and largely untouched by Western civilisation. There is some cultural variation, but by and large everyone is able to recognise the facial expressions, regardless of their origins. It seems our facial expressions are innate, rather than learned, 'hard-wired' into the human brain. Someone who grew up in the deepest recesses of the Amazon jungle would pull the same expression if something surprises them as someone who'd lived their entire life in New York.

Our brains are very adept at recognising and reading faces. Chapter 5 detailed how the visual cortex has subsections dedicated to processing faces, hence we tend to see them everywhere. So efficient is the brain in this regard that an expression can be deduced from minimal information, which is why it's common to now use basic punctuation to convey happiness :-) sadness :-(anger >:-(surprise :-O and many more. These are just simple lines and dots. They're not even upright. And yet we still perceive specific types of expression.

Facial expressions may seem a limited form of communication, but they're extremely useful. If everyone around you has a fearful expression, your brain instantly concludes there is something nearby that everyone considers a threat, and primes itself for fight or flight. If we had to rely on someone saying, 'I don't want to alarm you, but there appears to be a pack of rabid hyenas heading right for us', they'd probably be on us before the end of the sentence. Facial expressions also aid social interactions; if we're doing something and everyone has a happy expression, we know we should keep doing it to gain approval. If everyone looks at us and appears shocked, angry, disgusted or all three, then we should stop what we're doing rather quickly. This feedback helps guide our own behaviours.

Studies have revealed that the amygdala is highly active when we're reading facial expressions.[6] The amygdala, responsible for processing our own emotions, is seemingly necessary for recognising emotions in others. Other regions deep in the limbic system responsible for processing specific emotions (for instance, the putamen for disgust) are also implicated.

The link between emotions and facial expressions is strong but not insurmountable. Some people suppress or control their facial expressions so that they differ from their emotional state. The obvious example is the 'poker face'. Professional poker players maintain neutral expressions (or inaccurate ones) in order to hide how the cards dealt impact on their chances of winning. However, there is only a limited range of possibilities when being dealt cards from a deck of 52, and poker players can brace themselves for all of them, even an unbeatable straight flush. Knowing something is coming allows the more conscious controls of facial expressions to retain dominance. However, if during the game a meteorite crashes through the roof and onto the table, it's doubtful that any of the players could stop themselves from adopting a shocked expression.

This is indicative of yet another conflict between the advanced and primitive areas of the brain. Facial expressions can be voluntary (controlled by the motor cortex in the cerebrum) or involuntary (controlled by the deeper regions in the limbic system). Voluntary facial expressions we adopt by choice – for example, looking enthusiastic when viewing someone's tedious holiday photos. Involuntary expressions are produced by actual emotions. The advanced human neocortex may be capable of conveying inaccurate information (lying), but the older limbic control system is unfailingly honest, so they come into conflict quite often, because the norms of society often dictate that we don't give our honest opinion; if a person's new haircut repulses us, it's not done to say so.

Unfortunately, our brains being so sensitive to reading faces means we can often tell when someone is undergoing this internal conflict between honesty and manners (smiling through gritted teeth). Luckily, society has also deemed it impolite to point it out when someone is doing this, so a tense balance is achieved.

Carrots and sticks
(How the brain allows us to control others, and be controlled in turn)

I hate car shopping. Trudging across vast forecourts, checking endless details, looking at so many vehicles you lose all interest and start wondering if you have space in your garden for a horse. Feigning awareness of cars so you do things like kick the tyres. Why? Can the tip of your shoe analyse vulcanised rubber?

But for me, the worst part is car salesmen. I just can't deal with them. The machismo (I've yet to meet a female one), the exaggerated chumminess, the 'I'll have to ask the manager' tactic, the implication that they're losing money by my even being there. All these techniques confuse and unsettle me, and I find the whole process distressing.

That's why I always take my dad car shopping. He revels in this sort of thing. The first time he helped me buy a car I was braced for confident negotiating, but his tactic was largely swearing at the salesmen and calling them criminals until they agreed to lower the price. Unsubtle but definitely effective.

However, that car salesmen the world over have such established and recognisable methods suggests they do actually work. This is odd. All customers will have wildly different personalities, preferences and attention spans, so the idea that simple and familiar approaches will increase the odds of someone agreeing to hand over hard-earned cash should be ludicrous. However, there are specific behaviours that increase compliance, meaning customers agree with someone and 'submit to their will'.

We've covered how fear of social judgement causes anxiety; provocation triggers the anger system; and seeking approval can be a powerful motivator. Indeed, many emotions can be said to exist only in the context of other people: you can be angry at inanimate objects, but shame and pride require people's judgement, and love is something that exists between two people ('self-love' is something else entirely). So it's no great stretch to find that people can make others do what they want by exploiting the brain's tendencies. Anyone whose livelihood depends on convincing other people to give them money has familiar methods for increasing customer compliance and, once again, the way the brain works is largely responsible.

This doesn't mean there are techniques that give you total control over someone. People are far too complex, no matter what pick-up artists would have you believe. Nonetheless, there are some scientifically recognised means for getting people to comply with your wishes.

There's the 'foot-in-the-door' technique. A friend asks to borrow money for the bus. You agree. Then they ask if they can borrow more for a sandwich. You agree again. Then they say why not go to the pub, catch up over a few drinks? As long as you're OK to pay, they've not got any money, remember? You think, 'Sure, it's only a few drinks.' Then it's a few more and suddenly they're asking to borrow money for a taxi as they've missed the bus, and you sigh and agree because you've said yes to everything else.

If this so-called friend had said, 'Buy me dinner and drinks and pay for me to get home in a convenient manner', you'd have said no, because it's a ridiculous request. But that's exactly what you've done. This is the foot-in-the-door (FITD) technique, where agreeing with a small request will make you more amenable to a larger request. The requester has his 'foot in the door'.

FITD has several limitations, thankfully. There has to be a delay between the first and second request; if someone agrees to loan you £5, you can't ask for £50 ten seconds later. Studies have shown FITD can work days or weeks after the initial request, but eventually the association between the first and second requests is lost.

FITD also works better if requests are 'prosocial', something perceived as helpful, or doing good. Buying someone food is helpful, then loaning them money to get home is also helpful, so more likely to be a request that's complied with. Keeping lookout while someone scrawls obscenities on their ex's car is not good, so driving them to their ex's house to throw a brick through their window afterwards would be refused. Deep down, people are often quite nice.

FITD also needs consistency, for instance, loaning money,

then loaning more money. Driving someone home doesn't mean you'll look after their pet python for a month. How are these things related? Most people don't equate 'give a ride in my car' with 'have a giant snake in my house'.

Despite limitations, FITD is still potent. You've probably experienced the family member who gets you to set up a computer and ends up using you as 24/7 tech support, for instance. That's FITD.

A 2002 study by N. Guéguen shows it even works online.[7] Students who agreed with an emailed request to open a specific file were more likely to take part in a more demanding online survey when asked. Persuasion often relies on tone, presence, body language, eye contact and so on but this study shows these aren't necessary. The brain seems worryingly eager to agree with requests from people.

Another approach actually exploits a request that's been denied. Say someone asks you if they can store all their possessions in your house because they're moving out. This is inconvenient, so you decline. Then they ask if they can instead borrow your car for the weekend to move to their stuff elsewhere. This is a much easier, so you agree. But letting someone use your car for a weekend *is* inconvenient, just less so than the original request. Now you've got someone using your car, and you'd never usually agree to that.

This is the door-in-the-face technique (DITF). It sounds aggressive, but it's the person being manipulated who is 'slamming the door' into the face of those making demands. But slamming a door in someone's face (metaphorically or literally) makes you feel bad, so there's a desire to 'make it up' to them, hence agreeing with smaller requests.

DITF requests can be much closer together than FITD

ones; the first request is denied, so the person hasn't actually agreed to do anything yet. There is also evidence suggesting DITF is more potent. A 2011 study by Chan and her colleagues used FITD or DITF to compel groups of students to complete an arithmetic test.[8] FITD had a 60 per cent success rate, while DITF was closer to 90 per cent! The conclusion of this study was that if you want schoolchildren to do something, use a door-in-the-face approach, which is definitely something you should phrase differently when announcing it to the general public.

The potency and reliability of DITF may explain why it's so often used in financial transactions. Scientists have even assessed this directly: a 2008 study by Ebster and Neumayr[9] showed the DITF to be very effective when selling cheese from an Alpine hut to passers-by. (NB: Most experiments don't take place in Alpine huts.)

Then there's the low-ball technique, similar to FITD in that it results from someone initially agreeing to something, but which plays out differently.

Low-ball is where someone agrees to something (a specific price to pay, a certain amount of time to do a job, a specific word count for a document), then the other person suddenly increases the initial demand. Surprisingly, despite frustration and annoyance, most people will still agree with the increased demand. Technically, they have ample reason to refuse: it's someone breaking an agreement for personal gain. But people invariably comply with the suddenly increased demand, as long as it's not too excessive: if you agree £70 for a used DVD-player, you won't still agree if suddenly costs your life savings and firstborn child.

Low-ball can be used to make people *work for free*! Sort

of. A 2003 study by Burger and Cornelius of Santa Clara University had people agreeing to complete a survey in return for a free coffee mug.[10] They were then told there were no mugs available. Most still did the survey, despite not getting the promised reward. Another study by Cialdini and his colleagues in 1978 reported university students were far more likely to show up for a 7 a.m. experiment if they'd already agreed to show up at 9 a.m., than if they were initially asked to show up at 7 a.m.[11] Clearly, reward or cost aren't the only factors; many studies of the low-ball technique have shown that actively agreeing to a deal, willingly, before it's changed is integral to sticking to it regardless.

These are the more familiar of many approaches for manipulating people into complying with your wishes (another example is reverse psychology, which you definitely shouldn't look up yourself). Does this make much evolutionary sense? It's supposed to be 'survival of the fittest', but how is being easily manipulated a useful advantage? We'll look at this more in a later section, but the compliance techniques described here can all be explained by certain tendencies of the brain.*

A lot of these are linked to our self-image. Chapter 4 showed the brain (via the frontal lobes) is capable of self-analysis and awareness. So it's not so far-fetched that we'd use this information and 'adjust' for any personal failing. You've heard of people 'biting their tongue', but why do

* There's much theorising and speculation as to which brain processes and areas of the brain are responsible for these socially relevant tendencies, but it's difficult to pin these down even now. The more in-depth brain-scanning procedures such as MRI or EEG require the subject to be at least strapped into a large device in a lab, and it's difficult to get a realistic social interaction going in such contexts. If you were wedged into an MRI scanner and somebody you know wandered in and started asking you for favours, your brain would probably be more confused than anything.

that? They may think someone's baby is actually quite ugly, but stop themselves from saying this and instead say, 'Oh, how cute.' This makes people think better of them, whereas the truth wouldn't. This is something called 'impression management', which is where we try to control the impression people get of us via social behaviours. We care what other people think of us at a neurological level, and will go to great lengths to make them like us.

A 2014 study by Tom Farrow and his colleagues of the University of Sheffield suggested that impression management shows activation in the medial prefrontal cortex and left ventrolateral prefrontal cortex, along with other regions including the midbrain and cerebellum.[12] However, these areas were noticeably active only when subjects tried to make themselves look *bad*, when choosing behaviours to make people dislike them. If they were choosing behaviours that made them look *good*, there was no detectable difference from normal brain activity.

Coupled with the fact that subjects were much faster at processing behaviours that made them look good as opposed to bad, they concluded that making us look good to others is *what the brain is doing all the time*! Trying to scan for it is like trying to find a specific tree in a dense forest; there's nothing to make it stand out. The study in question was small, only 20 subjects, so it's possible specific processes for this behaviour might be found eventually, but the fact that there was still such a disparity between looking-good people and looking-bad people is striking.

But what does this have to do with manipulating people? Well, the brain seems to be geared towards making other people like it/you. All the compliance techniques arguably take advantage of a person's desire to be seen positively by others. This is such an ingrained drive that it can be exploited.

If you've agreed to a request, rejecting a similar request would probably cause disappointment and damage someone's opinion of you, so foot in the door works. If you've turned down a big request, you're aware that the person won't like you for this, so are primed to agree to a smaller request as a 'consolation', so door in the face works. If you've agreed to do or pay something and then the demand suddenly increases, backing out would again cause disappointment and make you look bad, so low ball works. All because we want people to think well of us, to the point where it overrides our better judgement or logic.

It's undoubtedly more complex than this. Our self-image requires consistency, so once the brain has made a decision it can be surprisingly hard to alter it, as anyone who's tried explaining to an elderly relative that not all foreigners are filthy thieves will know. We saw earlier how thinking one thing and doing something that contradicts it creates dissonance, a distressing state where thinking and behaviour don't match. In response, the brain will often alter its thinking to match the behaviour, restoring harmony.

Your friend wants money, you don't want to give it. But you just gave them a slightly smaller amount. Why would you do this if you didn't think it was acceptable? You want to be consistent, and liked, so your brain decides you *do* want to give them more money, and there we get the FITD. This also explains why making an active choice is important for low ball: the brain has made a decision, so will stick to it to be consistent, even if the reason for the decision no longer applies; you're committed, people are counting on you.

There's also the principle of reciprocity, a uniquely human phenomenon (as far as we know) where people will respond in kind to people being nice to them, more so than self-interest would suggest.[13] If you reject someone's request and they make a smaller one, you perceive this as them doing something nice for you, and agree to be disproportionately nice in turn. DITF is believed to exploit this tendency, because the brain interprets 'making a smaller request than the previous one' as someone doing *you* a favour, because it's an idiot.

As well as this, there's social dominance and control. Some (most?) people, in Western cultures at least, want to be seen as dominant and/or in control, because the brain sees this as a safer, more rewarding state. This can often manifest in questionable ways. If someone is asking you for things, they are subservient to you, and you stay dominant (and likeable) by helping them out. FITD fits nicely with this.

If you reject someone's request, you assert dominance, and if they make a smaller request they have established they're submissive, so agreeing with it means you can still be dominant and liked. A double whammy of good feelings. DITF can arise from this. And say you've decided to do something, then someone changes the parameters. If you then back out, this means *they* have control over *you*. To hell with that. You'll go through with the original decision anyway, because you're *nice,* damn it: low ball.

To summarise, our brains make us want to be liked, to be superior, and to be consistent. As a result of all this, our brains make us vulnerable to any unscrupulous person who wants our money and has a basic awareness of haggling. It takes an incredibly complex organ to do something this stupid.

Achy Breaky Brain
(Why a relationship break-up is so devastating)

Have you ever found yourself in the foetal position on the sofa for days on end, curtains drawn, phone unanswered, moving only to haphazardly wipe the snot and tears from your face, wondering why the very universe itself has cruelly decided to torment you so? Heartbreak can be all-consuming and totally debilitating. It is one of the most unpleasant things a modern human can expect to experience. It inspires great art and music as well as some terrible poetry. Technically, nothing has physically happened to you. You've not been injured. You've not contracted a vicious virus. All that's happened is you've been made aware that you won't be seeing a person you had a lot of interaction with much any more. That's it. So why does it leave you reeling for weeks, months, even for the rest of your life in some cases?

It's because other people have a major influence over our brain's (and therefore our) well-being, and seldom is this more true than in romantic relationships.

Much of human culture seems dedicated to ending up in a long-term relationship, or acknowledging that you're in one (see: Valentine's Day, weddings, rom-coms, love ballads, the jewellery industry, a decent percentage of all poetry, country music, anniversary cards, the game 'Mr & Mrs' and so on). Monogamy is not the norm among other primates[14] and seems odd when you consider we live much longer than the average ape so could feasibly dabble with many more partners in the available time. If it's all about 'survival of the fittest', making sure our genes propagate ahead of others, surely it would make more sense to reproduce with as many partners as possible, not stick to one person for our entire lives? But no, that's exactly what we humans tend to do.

There are numerous theories as to why humans are seemingly compelled to form monogamous romantic relationships, involving biology, culture, environment and evolution. Some argue that monogamous relationships result in two parents caring for offspring rather than one, so said offspring have greater chance of survival.[15] Others say it's due to more cultural influences, such as religion and class systems wanting to keep wealth and influence within the same narrow familial range (you can't make sure your family inherits your advantages if you can't keep track of it).[16] Another interesting new theory pins it on the influence of grandmothers acting as child carers, thus favouring the survival of long-term couples (even the most doting grandmother would probably baulk at caring for the unfamiliar offspring of her own child's ex).[17]

Whatever the initial cause, humans seem primed to seek out and form monogamous romantic relationships, and this is reflected in a number of weird things the brain does when we end up falling for someone.

Attraction is governed by many factors. Many species end up developing secondary sex characteristics, which are features that occur during sexual maturity but that aren't directly involved in the reproductive process, for instance, a moose's antlers or a peacock's tail. They're impressive and show how fit and healthy the individual creature is, but they don't *do* much beyond that. Humans are no different. As adults we develop many features that are apparently largely for physically attracting others: the deep voice, enlarged frames and facial hair of men, or the protruding breasts and pronounced curves of women. None of these things are 'essential', but in the distant past some of our ancestors decided that's what they wanted in a partner, and evolution took over from there. But then we end up with something of a chicken-and-egg scenario with regards to the brain, in that the human brain inherently finds certain features attractive *because it has evolved to do so.* Which came first, the attraction or the primitive brain's recognition of it? Hard to say.

Everyone has his or her own preferences and types, as we all know, but there are general patterns. Some of the things we humans find attractive are predictable, like the physical features alluded to above. Others are attracted to a more cerebral quality, with a person's wit or personality being the sexiest thing about them. A lot of variation is cultural, with what's deemed attractive being heavily influenced by things such as the media or what's considered 'different'. Contrast the popularity of false tans in more Western cultures with the huge market for body whitening lotions in many Asian countries. Some things are just bizarre, such as research that suggests people are more attracted to individuals that resemble them,[18] which harks back to the brain's ego bias.

It's important, however, to differentiate between a desire for sex, aka lust, and the deeper, more personal romantic attraction and bonding we associate with romance and love, things more often sought and found with long-term relationships. People can (and frequently do) enjoy purely physical sexual interactions with others that they have no real 'fondness' for apart from an appreciation for their appearance, and even that's not essential. Sex is a tricky thing to pin down with the brain as it underlies much of our adult thinking and behaviour. But this section isn't really about lust; we're talking more about *love*, in the romantic sense, for one specific individual.

There's a lot of evidence to suggest the brain does process these things differently. Studies by Bartels and Zeki suggest that when individuals who describe themselves as in love are shown images of their romantic partners, there is raised activity (not seen in lust or more platonic relationships) in a network of brain regions including the medial insula, anterior cingulate cortex, caudate nucleus and putamen. There was also *lower* activity in the posterior cingulate gyrus and in the amygdala. The posterior cingulate gyrus is often associated with painful emotion perception, so it makes sense that your loved one's presence would shut this down a bit. The amygdala processes emotions and memory, but often for *negative* things such as fear and anger, so again it makes sense that it's not so active now; people in committed relationships can often seem more relaxed and less bothered about day-to-day annoyances, regularly coming across as 'smug' to the independent observer. There's also diminished activity in regions including the prefrontal cortex, which is responsible for logic and rational decision-making.

Certain chemicals and transmitters are associated too.* Being in love seems to elevate dopamine activity in the reward pathway,[20] meaning we experience pleasure in our partner's presence, almost like a drug (see Chapter 8). And oxytocin

* One type of chemical often associated with attraction are pheromones, specific substances given off in sweat that other individuals detect and that alter their behaviour, most often linked with increasing arousal and attraction towards the source of the pheromones. While human pheromones are regularly referred to (you can seemingly buy sprays laced with them if you're looking to increase your sexual appeal), there's currently no definitive evidence that humans have specific pheromones that influence attraction and arousal.[19] The brain may often be an idiot, but it's not so easily manipulated.

is often referred to as 'the love hormone' or similar, which is a ridiculous oversimplification of a complex substance, but it does seem to be elevated in people in relationships, and it has been linked to feelings of trust and connection in humans.[21]

This just the raw biological stuff that happens in our brains when we fall in love. There's also many other things to consider, like the expanded sense of self and achievement that comes from being in a relationship. There's the immense satisfaction and achievement that comes from having a whole other person value you so highly and want to be in your company in all manner of contexts. Given that most cultures invariably see being in a relationship as a universal goal or achievement (as any happily single person will tell you, usually through gritted teeth), there's also advanced social standing from being in a couple.

The flexibility of the brain also means that, in response to all this deep and intense stuff that results from being committed to someone, it adapts to expect it. Our partners become integrated into our long-term plans, goals and ambitions, our predictions and schemas, our general way of thinking about the world. They are, in every sense, a big part of our life.

And then it ends. Maybe one partner wasn't being faithful; maybe there's just not enough compatibility; perhaps one partner's behaviour drove the other away. (Studies have shown that people with more anxious tendencies tend to exaggerate and amplify relationship conflicts, possibly to breaking point.[22])

Consider everything the brain invests in sustaining a relationship, all the changes it undergoes, all the value it places on being in one, all the long-term plans it makes, all the familiar routines it grows to expect. If you remove all this in one fell swoop, the brain is going to be seriously negatively affected by it.

All the positive sensations it has grown to expect suddenly cease. Our plans for the future and expectations of the world are suddenly no longer valid, which is incredibly distressing for an organ that, as we've seen repeatedly, doesn't deal with uncertainty and ambiguity well at all. (Chapter 8 goes into all of this in more detail.) And there is copious practical uncertainty to deal with if it was a long-term relationship. Where will you live? Will you lose your friends? What about the financial concerns?

The social element is also quite damaging, considering how much we value our social acceptance and standing. Having to explain to all your friends and family that you 'failed' at a relationship is bad enough, but consider the break-up itself; someone who knows you better than anyone, at the most intimate level, has deemed you unacceptable. This is a real kick in the social identity. This is where it hurts.

That's a literal comment by the way; studies have shown that a relationship break-up activates the same brain regions that process physical pain.[23] There have been numerous examples throughout this book of the brain processing social concerns in the same way as genuine physical concerns (for example, social fears being just as unnerving as actual physical danger), and this is no different. They say 'love hurts', and, yes, yes it does. Paracetamol is even sometimes effective for 'heartache'.

Add to this that you have countless memories of that person that were formally happy but that are now linked with something very negative. This undermines a big part of your sense of self. And, on top of that, the earlier observation that being in love is like a drug comes back to haunt you; you're used to experiencing something constantly rewarding, and suddenly it's taken away. In Chapter 8, we'll see how addiction and withdrawal can be very disruptive and damaging to the brain, and a not dissimilar process is happening here when we experience a sudden break-up with a long-term partner.[24]

This isn't to say the brain doesn't have the ability to deal with a break-up. It can put everything back together eventually, even if it's a slow process. Some experiments showed that specifically focusing on the positive outcomes of a break-up can cause more rapid recovery and growth,[25] as alluded to earlier in the brain's bias for preferring to remember 'good' things. And, just sometimes, science and clichés match up, and things really do get better with time.[26]

But overall, the brain dedicates so much to establishing and sustaining a relationship that it suffers, as do we, when it all comes crashing down. 'Breaking up is hard to do' is an understatement.

People power
(How the brain reacts to being part of a group)

What exactly is a 'friend'? It's a question that makes you seem a rather tragic individual if asked aloud. A friend is essentially someone with whom you share a personal bond (that isn't familial or romantic). However, it's more complicated because people have many different categories of friends; work friends, school friends, old friends, acquaintances, friends you don't really like but have known too long to get rid of, and so on. The Internet also now allows 'online' friends, as people can form meaningful relationships with like-minded strangers across the planet.

It's lucky we have powerful brains, capable of handling all these different relationships. Actually, according to some scientists, this isn't just a convenient coincidence; we may have big powerful brains *because* we formed complicated social relationships.

This is the social brain hypothesis, which argues that complex human brains are a result of human friendliness.[27] Many species form large groups, but this doesn't equal intelligence. Sheep form flocks, but their existence seems is largely dedicated to eating grass and general fleeing. You don't need smarts for that.

Hunting in packs requires more intelligence as it involves coordinated behaviours, so pack hunters such as wolves tend to be smarter than docile-but-numerous prey. Early human communities were substantially more complex again. Some humans hunt, while others stay and look after the young and sick, protect the homestead, forage for food, make tools and so on. This cooperation and division of labour provides a safer environment all round, so the species survives and thrives.

This arrangement requires humans to care about others *who are not biologically linked to them*. It goes beyond simple 'protect our genes' instincts. Thus, we form friendships, meaning we care about the well-being of others even though our only biological connection is that we're the same species (and 'man's best friend' shows even this isn't essential).

Coordinating all the social relationships required for community life demands a great deal of information processing. If pack hunters are playing noughts and crosses, human communities are engaged in ongoing chess tournaments. Consequently, powerful brains are needed.

Human evolution is difficult to study directly, unless you

have several hundred thousand years to spare and *lots* of patience, so it's hard to determine the accuracy of the social-brain hypothesis. A 2013 Oxford University study claimed to have demonstrated it via sophisticated computer models that showed social relationships do in fact require more processing (and therefore brain) power.[28] Interesting, but not conclusive; how do you model friendship on a computer? Humans have a strong tendency to form groups and relationships, and concern for others. Even now, a complete lack of concern or compassion is considered abnormal (psychopathy).

An inherent tendency to want to belong to a group can be useful for survival, but it also throws up some surreal and bizarre results. For example, being part of a group can override our judgement, even our senses.

Everyone knows about peer pressure, where you do or say things not because you agree but because the group you belong to wants you to, like claiming to like a band you detest because the 'cool' kids like them, or spending hours discussing the merits of a film your friends loved but that you found agonisingly dull. This is a scientifically recognised occurrence, known as normative social influence, which is what happens when your brain goes to the effort of forming a conclusion or opinion about something, only to abandon it if the group you identify with disagrees. Worryingly often, our brains prioritise 'being liked' over 'being right'.

This has been demonstrated in scientific settings. A 1951 study by Solomon Asch put subjects in small groups and asked them very basic questions; for instance, showing them three different lines and asking, 'Which is longest?'[29] It might surprise you to hear that most participants gave completely the wrong answer. It didn't surprise the researchers though, because only one person in each group was a 'real' subject; the rest were stooges instructed to give the wrong answer. The genuine subjects had to give their answers last, when everyone given theirs aloud. And 75 per cent of the time, the subjects gave the wrong answer too.

When asked why they gave a clearly wrong answer, most said they didn't want to 'rock the boat' or similar sentiments. They didn't 'know' the other group members at all outside the experiment, and yet they wanted the approval of their new peers, enough to deny their own senses. Being part of a group is apparently something our brains prioritise.

It's not absolute. Although 75 per cent of subjects agreed with the group's wrong answer, 25 per cent *didn't*. We may be heavily influenced by our group but our own backgrounds and personalities are often equally potent, and groups are composed of different types of individuals, not submissive drones. You do get people who are happy to say things almost everyone around them will object to. You can make millions doing this on televised talent shows.

Normative social influence can be described as behavioural in nature; we *act* as if we agree with the group, even if we don't. The people around us can't dictate how we *think* though, surely?

Often, this is true. If all your friends and family suddenly insisted 2 + 2 = 7, or that gravity pushes you up, you still wouldn't agree. You might worry that everyone you care about has completely lost it, but you wouldn't agree, because your own senses and understanding show that they're wrong. But here the truth is blatant. In more ambiguous situations, other people can indeed impact on our thought processes.

This is informational social influence, where other people

are used by our brains as a reliable source of information (however wrongly) when figuring out uncertain scenarios. This may explain why anecdotal evidence can be so persuasive. Finding accurate data about a complex subject is hard work, but if you heard it from a bloke down the pub, or from your friend's mother's cousin who knows about it, then this is often sufficient evidence. Alternative medicine and conspiracy theories persist thanks to this.

It's perhaps predictable. For a developing brain, the main source of information is other people. Mimicry and imitation are fundamental processes whereby children learn, and for many years now neuroscientists have been excited about 'mirror neurons', neurons that activate both when we perform a specific action and when we observe that action from someone else, suggesting the brain recognises and processes the behaviour of others at a fundamental level. (Mirror neurons and their properties are something of a controversial issue in neuroscience, so don't take any of this for granted.[30])

Our brains prefer to use other people as a go-to reference for information in uncertain scenarios. The human brain evolved over millions of years, and our fellow humans have been around a lot longer than Google. You can see how this would be useful; you hear a loud noise and think it might be an enraged mammoth, but everyone else in your tribe is running away screaming, so they probably know it *is* an enraged mammoth, and you'd best follow suit. But there are times when basing your decisions and actions on other people's can have dark and unpleasant consequences.

In 1964, New York resident Kitty Genovese was brutally murdered. While tragic in itself, this particular crime became infamous because reports revealed that 38 people witnessed the attack but did nothing to help or intervene. This shocking behaviour prompted social psychologists Darley and Latané to investigate it, leading to the discovery of the phenomenon known as the 'bystander effect',* which is where people are unlikely to intervene or offer assistance if there are others around.[31] This isn't (always) due to selfishness or cowardice but because we refer to other people to determine our actions when we aren't certain what to do. There are plenty of people who get stuck in where needed, but if others are around the bystander effect presents a psychological obstacle that must be overcome.

The bystander effect acts to suppress our actions and decisions; it stops us doing something because we're in a group. Being part of a group can also cause us to think and do things we'd never do when alone.

Being in a group invariably makes people desire group harmony. A fractious or argumentative group isn't useful and is unpleasant to be part of, so overall agreement and accord is usually something everyone wants to achieve. If conditions are right, this desire for harmony can be so compelling that people will end up thinking or agreeing with things that they'd usually consider irrational or unwise just to achieve it. When the good of the group takes precedence over logical or reasonable decisions, this is known as groupthink.[32]

Groupthink is only part of it. Take a controversial subject matter, like the legalisation of cannabis (something that's a

* Retrospective investigations suggest the original reports of the crime were inaccurate, more urban legend than accurate report, something made up to sell newspapers. Despite this, the bystander effect is a real phenomenon. The murder of Kitty Genovese and supposed unwillingness of witnesses to intervene had other surreal consequences; it's referenced in Alan Moore's ground-breaking comic *Watchmen*, as the event that leads to the character Rorschach taking up vigilantism. Many say they'd love superhero comics to be real. Be careful what you wish for.

'hot button' issue at the time of writing). If you took 30 people off the street (with their permission) and asked them their thoughts about legalising cannabis, you'd likely get a range of opinions, from, 'Cannabis is evil and you should be locked up for even smelling it', to, 'Cannabis is great and should be given away with children's meals', with most falling somewhere between these two extremes.

If you put these people together in one group and ask them to come up with a consensus on cannabis legalisation, you'd logically expect something that is the 'average' of everyone's individual opinion, such as: 'Cannabis shouldn't be legalised but possession should only be a minor offence.' But, as ever, logic and brain don't really see eye to eye. Groups will often adopt a *more* extreme conclusion than individual members would if alone.

Groupthink is part of it, but we also want to be liked by the group, and achieve high status in it. So Groupthink produces a consensus that members agree with, but they'll also agree with it more strongly, to impress the group. But then others do that too, and everyone ends up trying to outdo each other.

'So we agree cannabis shouldn't be legalised. Possession of any amount of it should be an arrestable offence.'

'Arrestable? No, guaranteed jail, ten years for possession!'

'Ten years? I say life imprisonment!'

'Life? You hippy! A death sentence, at the very least.'

This phenomenon is known as group polarisation, where people in groups end up expressing views that are more extreme than those they have when alone.* It's very common

* Fans of *Monty Python* should be familiar with the 'Four Yorkshiremen' sketch. This is (presumably accidentally) an excellent example of group polarisation, if a rather surreal one by normal standards.

and warps group decision-making in countless circumstances. It can be limited or prevented by allowing criticism and/or outside opinions to be aired, but the powerful desire for group harmony usually prevents this by excluding detractors and rational analysis from discussions. This is alarming, because countless decisions that affect millions of lives are made by like-minded groups who don't allow outside input. Governments, the military, corporate boardrooms – what makes these immune to making ridiculous conclusions resulting from group polarisation?

Nothing, nothing at all. A lot of the baffling or worrying policies pursued by governments could be explained by group polarisation.

Bad decisions by the powerful often result in angry mobs, another example of the alarming effects being part of a group can have on the brain. People are very good at perceiving the emotional states of others; if you've ever wandered into a room where a couple have just had a row, you can palpably feel the 'tense atmosphere' even though nobody is saying anything. This isn't telepathy or anything 'sci-fi', just our brains being attuned to picking up this sort of information through various cues. But when surrounded by people in the same intense emotional state, this can heavily influence our own, hence we're far more likely to laugh when part of an audience. As always, this can go too far.

Under certain conditions, the highly emotional or aroused state of those around us actually suppresses our individuality. We need a dense or closely unified group that allows us anonymity, that is highly aroused (experiencing strong emotions, not . . . something seedier), and with a focus on external events, so as to avoid thinking about the group's actions. Angry mobs and riots are perfect for creating these circumstances, and when these conditions are met we undergo a process known as 'deindividuation',[33] which is the scientific term for 'mob mentality'.

With deindividuation, we lose our usual ability to suppress impulses and think rationally; we become more prone to detecting and responding to the emotional states of others, but lose our typical concern for being judged by them. These in conjunction make people behave very destructively when part of a mob. Exactly how or why is difficult to say; it's hard to study this process scientifically. You rarely get an angry mob in a laboratory unless they've heard about your grave robbing and are there to put an end to your ungodly efforts to raise the dead.

I'm not mean, but my brain is
(The neurological properties that make us treat others badly)

Thus far, it seems the human brain is geared towards forming relationships and communicating. Our world should be nothing but people holding hands, singing happy songs about rainbows and ice-cream. However, human beings are frequently *terrible* to each other. Violence, theft, exploitation, sexual assault, imprisonment, torture, murder – these aren't rare; your typical politician has probably indulged in many. Even genocide, attempting to wipe out an entire population or race, is familiar enough to warrant a dedicated term.

Edmund Burke famously said, 'The only thing necessary for the triumph of evil is for good men to do nothing.' But it's probably even easier for evil if good men are willing to pitch in and help.

But *why* would they do it? There are numerous explanations regarding cultural, environmental, political, historical factors, but the workings of the brain also contribute. At the Nuremberg trials, where those responsible for the Holocaust were questioned, the most common defence was they were 'only following orders'. A feeble excuse, right? Surely, no normal person would do such awful things, no matter who told them to? But, alarmingly, it seems they just might.

Stanley Milgarm, a Yale professor, studied this 'only obeying orders' claim in an infamous experiment. It involved two subjects, in separate rooms, where one had to ask the other questions. If a wrong answer was given, the questioner had to administer an electric shock. For every wrong answer, the voltage was increased.[34] Here's the catch: there were no shocks. The subject answering questions was an actor, deliberately getting things wrong and giving increasingly pained sounds of distress whenever a 'shock' was delivered.

The real subject of the experiment was the questioner. The set-up meant they believed they were essentially torturing a person. Subjects invariably showed discomfort or distress over this, and objected or asked to stop. The experimenter always said the experiment was important so they must continue. Disconcertingly, 65 per cent of people did, continuing to inflict intense pain on someone purely because they were told to.

The researchers didn't trawl the maximum security cells of prisons for volunteers; everyone who took part was a normal everyday person, who was surprisingly willing to torture another person. They might have objected to it, but they still *did it*, which is the more relevant point for the recipient.

This study has had numerous follow-ups that provide more specific information.* People were more obedient if the experimenter was in the room, rather than communicating via telephone. If subjects saw other 'subjects' refuse to obey, they were more likely to disobey themselves, suggesting that people are willing to be rebels, just not the *first* rebel. Experimenters wearing lab coats and conducting the experiments in professional-looking offices also increased obedience.

The consensus is that we're willing to obey *legitimate* authority figures, who are seen as responsible for consequences of actions they demand. A remote person who is visibly disobeyed is harder to consider authoritative. Milgram proposed that, in social situations, our brains adopt one of two states: an autonomous state (where we make our own decisions) and an agentic state, where we allow others to dictate our actions, although this hasn't yet been reliably identified in any brain-scanning studies.

One idea is that, in evolutionary terms, a tendency to obey unthinkingly is more efficient; stopping to fight about who's in charge every time a decision needs to be made is very impractical, so we're left with a tendency to obey authority despite any reservations. It's no great stretch to imagine corrupt but charismatic leaders exploiting this.

However, people are regularly horrible to others without orders from some tyrannical authority. Often it's one group of people making life miserable for another, for various reasons.

* There have also been many criticisms of these experiments. Some are to do with methods and interpretations, whereas others are about ethics. What right have scientists to make innocent people think they are torturing others? Such realisations can be very traumatic. Scientists have a reputation for being cold and dispassionate, and it's sometimes easy to see why.

The 'group' element is important. Our brains compel us to form groups, and turn on those who threaten them.

Scientists have studied what it is about the brain that makes us so hostile to anyone who dares disrupt our group. One study by Morrison, Decety and Molenberghs suggested that when subjects contemplate being part of a group, the brain shows activation in a neural network composed of cortical midline structures, tempo-parietal junctions and anterior temporal gyrus.[35] These regions have been shown repeatedly to be highly active in contexts where interaction and thinking of others is required, meaning some have dubbed this particular network the 'social brain'.*[36]

Another particularly intriguing finding was that when subjects had to process stimuli that involved being part of a group, activity was seen in a network including the ventral medial prefrontal and anterior and dorsal cingulate cortex. Other studies have linked these areas to processing of the 'personal self',[37] suggesting considerable overlap between self-perception and group membership. This means people derive much of their identity from the groups they belong to.

One implication of this is that any threat to our group is essentially a threat to 'ourselves', which explains why anything that poses a danger to our group's way of doing things is met with such hostility. And the main threat to most groups are . . . other groups.

Fans of rival football teams engage in violent clashes so often they're practically a continuation of the actual game. Warfare between rival criminal gangs is a staple of gritty crime dramas. Any modern political contest quickly becomes

* Not to be confused with the social-brain hypothesis from earlier, because scientists never miss an opportunity to be confusing.

a battle between one side and another, where attacking the opposition is more important than explaining why anyone should vote for you. The Internet has just made things worse: post even a slightly critical or controversial opinion online about anything anyone finds important (for example, the *Star Wars* prequels weren't that bad, actually) and you'll have an inbox clogged with hate mail before you've can put the kettle on. I write blogs for an international media platform, so trust me on this.

Some may think prejudices come from long periods of exposure to the attitudes that shape them; we aren't born with an inherent dislike of certain types of people, it must need the slow drip-drip of (metaphorical) bile over the years to wear down someone's principles and make them hate others unreasonably. That is often true. It can also happen very quickly.

The infamous Stanford Prison experiment, run by a team lead by Philip Zimbardo, looked at the psychological consequences of the prison environment on guards and prisoners.[38] A realistic prison set was constructed in the Stanford University basements, and subjects were designated either prisoners or guards.

The guards became incredibly cruel, being rude, aggressive, abusive and hostile to prisoners. The prisoners ended up thinking of the guards (quite reasonably) as unhinged sadists, so they organised a rebellion, barricading themselves in their rooms, which guards stormed and stripped. Prisoners soon became prone to depression, sobbing fits, even psychosomatic rashes.

The duration of the experiment? Six days. It was planned for two weeks, but was halted early because things got so bad. It's important to remember *none of them were really prisoners or guards*! They were students, from a prestigious university. But they were placed in clearly identified groups, made to coexist with another group with different goals, and group mentality exerted itself very quickly. Our brains are very quick to identify with a group, and in certain contexts this can seriously alter our behaviour.

Our brain makes us hostile to those who 'threaten' our group, even if it's a trivial matter. Most of us know this from schooldays. Some unfortunate individual inadvertently does something that deviates from the group's normal standards of behaviour (gets an unusual haircut), which undermines the group uniformity, and is punished (endlessly mocked).

Humans don't just want to be part of a group; they want a high-ranking role in it. Social status and hierarchy is very common in nature; even chickens have a hierarchy – hence the term 'pecking order' – and humans are just as keen on enhancing their social status as the proudest chicken – hence the term 'social climber'. They try to outdo each other, make themselves look good/better, be the comparative best at what they do. The brain facilitates this behaviour via regions including the inferior parietal lobe, dorsolateral and ventrolateral prefrontal cortices, fusiform and lingual gyri. These areas collaborate to provide awareness of social standing, so that we're not only aware of our membership of a group, but of our position in it.

As a result, anyone who does something that doesn't meet the group's approval is both risking the 'integrity' of the group and presenting an opportunity for other members to increase their status at the incompetent individual's expense. Hence, name calling and mockery.

However, the human brain is so sophisticated that the 'group' we belong to is a very flexible concept. It can be an entire country, as anyone waving their national flag demonstrates. People can even feel like a 'member' of a specific race, which is arguably easier as race stems from certain physical characteristics, so members of other races are easily identified and attacked by those who have so little to be proud of that their physical traits (which they had no role in obtaining) are very precious to them.

Disclaimer: I'm not a fan of racism.

But there are times when humans, individually, can be alarmingly cruel to those who don't deserve it. The homeless and poor, victims of assault, the disabled and sick, desperate refugees; rather than getting much needed help, these people are vilified by those better off. This goes against every facet of human decency and basic logic. So why's it so common?

The brain has a strong egocentric bias; it makes it and us look good at every opportunity. This can mean that we struggle to empathise with people – because they aren't us – and the brain mostly has things that have happened to us to go on when making decisions. However, a part of the brain, mainly the right supramarginal gyrus, has been shown to recognise and 'correct' this bias, allowing us to empathise properly.

There's also data showing it's much harder to empathise when this area is disrupted, or you aren't given time to think about it. Another intriguing experiment, lead by Tania Singer from the Max Planck Institute, showed that there are other limits to this compensatory mechanism, by exposing pairs of people to varying tactile surfaces (they had to touch either something nice or something gross).[39]

They showed two people experiencing something unpleasant will be very good at empathising correctly, recognising the emotion and intensity of feeling in the other person, but if one is experiencing pleasure while the other is enduring unpleasantness, then the pleasure-experiencing person will seriously underestimate the other's suffering. So the more privileged and comfortable someone's life is, the harder it is for them to appreciate the needs and issues of those worse off. But as long as we don't do something stupid like put the most pampered people in charge of running countries, we should be OK.

We have seen that the brain has an egocentric bias. Another (related) cognitive bias is called the 'just world' hypothesis.[40] This argues that the brain inherently assumes the world is fair and just, where good behaviour gets rewarded and bad behaviour is punished. This bias helps people function as a community because it means bad behaviour is deterred before it happens, and people are inclined towards being nice (not that they wouldn't be anyway, but this helps). It also motivates us; believing the world is random and all actions are ultimately meaningless won't help you get out of bed at a reasonable hour.

Unfortunately, this isn't true. Bad behaviour isn't always punished; good people often have bad things happen to them. But the bias is so ingrained in our brains that we stick to it anyway. So when we see someone who is an undeserving victim of something awful, this sets up a dissonance: the world is fair, but what happened to this person isn't fair. The brain doesn't like dissonance, so has two options: we can conclude the world is cruel and random after all, or decide that the victim *did something to deserve it*. The latter is crueller, but it lets us keep our nice cosy (incorrect) assumptions about the world, so we blame victims for their misfortune.

Numerous studies have shown this effect and its many manifestations. For example, people are less critical of victims if they themselves can intervene to alleviate their suffering, or if they were told the victims were compensated later. If people have no means to help victims, they'll be more disparaging towards them. This, while seeming especially harsh, is consistent with the 'just world' hypothesis: the victims have no positive outcome, so they *must* deserve it, surely?

People are also far more likely to blame a victim they strongly identify with. If you see someone of a different age/race/gender get hit by a falling tree, it's much easier to sympathise. But if you see someone of your age, height, build, gender, driving a car just like yours and colliding with a house like the one you live in, you're far more likely to blame that someone for being incompetent or stupid, despite having no evidence of this.

In the first instance, none of the factors apply to us, so it's OK to blame random chance for what happens; it's something that can't affect us. The second could easily apply to us, so the brain rationalises it as the fault of the individual involved. It must be *their* fault, because if it was random chance then it could happen to *you*. And that's upsetting thinking.

It seems that, despite all the inclinations towards being sociable and friendly, our brain is so concerned with preserving a sense of identity and peace of mind that it makes us willing to screw over anyone and anything that could endanger this. Charming.

8

When the brain breaks down . . .

Mental health problems, and how they come about

What have we learned so far about the human brain? It messes about with memories, it jumps at shadows, it's terrified of harmless things, it screws with our diet, our sleeping, our movement, it convinces us we're brilliant when we're not, it makes up half the stuff we perceive, it gets us to do irrational things when emotional, it causes us to make friends incredibly quickly and turn on them in an instant.

A worrying list. What's even more worrying, it does all of this *when it's working properly*. So what happens when the brain starts to go, for want of a better word, wrong? That's when we can end up with a neurological or mental disorder.

Neurological disorders are due to physical problems or disruption in the central nervous system, like damage to the hippocampus causing amnesia or degradation of the substantia nigra leading to Parkinson's disease. These things are awful, but usually have identifiable physical causes (although we often can't do much about them). They mostly manifest as physical issues, like seizures, movement disorders, or pain (migraines, for example).

Mental disorders are abnormalities of thinking, behaviour or feeling, and they need not have clear 'physical' cause. Whatever's causing them is still based in the physical make-up of the brain, but the brain is physically normal; it's just doing unhelpful things. To invoke the dubious computer analogy again, a neurological disorder is a hardware problem, whereas a mental disorder is a software problem (although there's ample overlap between the two, it's nowhere near as clear cut).

How do we define a mental disorder? The brain is made up of billions of neurons forming trillions of connections producing thousands of functions derived from countless genetic processes and learned experiences. No two are exactly alike, so how do we determine whose brain is working normally and whose 'isn't'? Everyone has weird habits, quirks, tics or eccentricities, which are often incorporated into identity and personality. Synaesthesia, for instance, doesn't seem to cause anyone any problems with functioning; many people don't realise they have anything amiss until they get weird looks for saying they like the smell of purple.[1]

Mental disorders are generally described as patterns of behaviour or thinking that cause discomfort and suffering, or impaired ability to function in 'normal' society. That last bit is important; it means for a mental disorder to be recognised it has to be compared with what's 'normal', and this can vary considerably over time. Only in 1973 did the American Psychiatric Association declassify homosexuality as a mental disorder.

Mental health practitioners are constantly revaluating the categorisation of mental disorders due to advances in understanding, new therapies and approaches, changes in dominant schools of thought, even the worrying influence of pharmaceutical companies, who like having new ailments to sell medications for. This is all possible because, up close, the line between 'mental disorder' and 'mentally normal' is incredibly fuzzy and indistinct, often relying on arbitrary decisions based on social norms.

Add to this the fact they're so common (nearly 1 in 4 people experience some manifestation of mental disorder, according to the data[2]) and it's easy to see why mental health problems are such a controversial issue. Even when they are recognised as a real thing (which is far from a given), the debilitating nature of mental disorders is often dismissed or ignored by those lucky enough not to be afflicted. There is also heated debate about how to classify mental disorders. For example, many say 'mental illness', but there are those who find this term misleading; it implies something that can be remedied, like the flu, or chickenpox. Mental disorders don't work that way; there often isn't a physical problem to be 'fixed', meaning a 'cure' is hard to identify.

Some even strongly object to the term 'mental disorder' as it makes them seem bad or damaging, when they can instead be seen as alternative ways of thinking and behaving. There's a large swathe of the clinical psychology community who argue that talking and thinking of mental issues as illnesses or problems is itself harmful, and are pushing for more neutral and less loaded terms to be used when discussing them. There are growing objections to the dominance of the medical field and approaches to mental health, and given the arbitrary nature of establishing what's 'normal' or not, this is understandable.

Despite these arguments, this chapter does stick more to the medical/psychiatric perspective – that's my background and, for most of us, it's the most familiar way of describing the subject matter. This is a brief overview of some more common examples of mental health issues while explaining how our brains let us down, both for those afflicted by the problem, and those of us around them who so often struggle to recognise and appreciate what's going on.

Dealing with the black dog
(Depression and the misconceptions around it)

Depression, the clinical condition, could use a different name. 'Depressed' presently applies both to people who are a bit miserable and to those with a genuine debilitating mood disorder. This means people can dismiss depression as a minor concern. After all, everyone gets depressed now and again, right? We just get over it. We often have only our own experiences to base judgements on, and we've seen how our brains automatically big up and exaggerate our own experiences, or minimise our impression of other people's experiences if they differ from our own.

This doesn't make it right, though. Dismissing the concerns of a person with genuine depression because you've been miserable and got over it is like dismissing someone who's had to have their arm amputated because you once had a papercut. Depression is a genuine debilitating condition, and being in 'a bit of a funk' isn't. Depression can be so bad that those experiencing it end up concluding that ending their life is the only viable option.

It's an indisputable fact that everyone dies eventually. But knowing it and directly experiencing it are two different things; you can 'know' that it hurts to get shot, but this doesn't mean you know how getting shot feels. Similarly, we know that everyone close to us will expire eventually, but it's still an emotional gut punch when it happens. We've seen how the brain has evolved to form strong and lasting relationships with people, but the down side is how much it hurts when those relationships come to an end. And there's no 'end' more final than when someone dies.

As bad as this is, there's an extra dose of awfulness when a loved one ends their own life. How and why someone ends up believing suicide is the only viable option is impossible for us to know for certain, but whatever the reasoning it's devastating to those left behind. These people are the ones the rest of us get to see. As a result, it's understandable why people often form negative opinions of the deceased – they might have successfully ended their own suffering, but they've caused it in many others.

As we saw in Chapter 7, the brain performs serious mental gymnastics to avoid feeling sorry for victims, and another possible manifestation of this is the labelling of those who end their own lives as 'selfish'. It's a bitterly ironic coincidence that one of the most common factors leading to suicide is clinical depression, as people with it are also regularly labelled as 'selfish', 'lazy' or with other disparaging adjectives. This may be the brain's egocentric self-defence kicking in again; acknowledging a mood disorder so severe that ending it all is an acceptable solution technically means acknowledging, at some level, that it might happen to you. An unpleasant thought. But if someone's just self-indulgent or callously selfish, that's their problem. It won't happen to *you*, and thus you get to feel better about yourself.

That's one explanation. Another is that some people are just ignorant jerks.

Labelling those with depression and/or those who die by suicide as selfish is a bleakly common occurrence, most prominently seen when applied to someone even slightly famous. The sad passing of Robin Williams, international superstar and beloved actor and comedian, provides the most obvious recent example.

Amid the glowing and tearful tributes, the media and Internet were still awash with comments like, 'Doing that to your family is just selfish', or, 'To commit suicide when you've got so much going for you is pure selfishness', and so on. These comments weren't restricted to anonymous online types; such sentiments came from high-profile celebrities and numerous news networks not exactly known for compassion, such as Fox News.

If you are someone who has expressed these views or similar, sorry – but you're wrong. Quirks of the brain's workings may explain part of it, but ignorance and misinformation can't be ignored. Granted, our brains don't like uncertainty and unpleasantness, but most mental disorders provide ample amounts of both. Depression is a genuine and serious problem that deserves empathy and respect, not dismissal and scorn.

Depression manifests in many different ways. It's a mood disorder, so mood is affected, but *how* it's affected varies. Some end up with unshakeable despair; others experience intense anxiety, resulting in feelings of impending doom and alarm. Other people have no mood to speak of, just feel empty and emotionless regardless of what's happening. Some (mostly men) become constantly angry and restless.

This is part of why it's proven difficult to establish an underlying cause of depression. For some time, the most widespread theory was the monoamine hypothesis.[3] Many neurotransmitters used by the brain are types of monoamines, and people with depression seem to have reduced levels of them. This affects the brain's activity, in a manner that may lead to depression. Most well-known antidepressants increase the availability of monoamines in the brain. The currently most widely used antidepressants are selective serotonin reuptake inhibitors (SSRI). Serotonin (a monoamine) is a neurotransmitter involved in processing anxiety, mood, sleep and so on. It's also believed to help regulate other neurotransmitter systems, so altering its levels could have a 'knock-on' effect. SSRIs work by stopping the removal of serotonin from synapses after it's released, increasing overall levels. Other antidepressants do similar things with monoamines such as dopamine or noradrenaline.

However, the monoamine hypothesis is meeting increasing criticism. It doesn't really explain what's happening; it's like restoring an old painting and saying it 'needs more green'; that might well be the case, but it's not specific enough to tell you what you actually need to do.

Also, SSRIs raise serotonin levels immediately, but beneficial effects take weeks to be felt. Exactly why this is has yet to be established (although there are theories, as we'll see), but it's like filling your car's empty tank with petrol and it working again only a month later; 'no fuel' may have been a problem, but it's clearly not the *only* problem. Add to this the lack of evidence showing a specific monoamine system that's impaired in depression, and that some effective antidepressants which don't interact with monoamines at all, and clearly there's more to depression than a simple chemical imbalance.

Other possibilities abound. Sleep and depression also seem

interlinked[4] – serotonin is a key neurotransmitter in regulating circadian rhythms, and depression causes disturbed sleep patterns. The first chapter showed sleep disruption is problematic; maybe depression is another consequence?

The anterior cingulate cortex has also been implicated in depression.[5] It's a part of the frontal lobe that seems to have many functions, from monitoring heart rate to anticipating reward, decision-making, empathy, controlling impulses and so on. It's essentially a cerebral Swiss Army knife. It's also been shown to be more active in depressed patients. One explanation is it's responsible for cognitive experience of suffering. If it is responsible for anticipation of reward then it makes sense that it would be involved in perceiving pleasure or, more pertinently, a complete lack thereof.

The hypothalamic axis that regulates responses to stress is also a focus of study.[6] But other theories suggest that the mechanism of depression is more of a widespread process than being isolated in specific brain areas. Neuroplasticity, the ability to form new physical connections between neurons, underpins learning and much of the brain's general functioning, and has been shown to be impaired in people with depression.[7] This arguably prevents the brain from responding or adapting to aversive stimuli and stress. Something bad happens, and the impaired plasticity means the brain is more 'fixed', like a cake left out too long, preventing moving on or escaping the negative mind-set. Thus, depression happens and endures. This might explain why depression is so persistent and pervasive; impaired neuroplasticity prevents a coping response. Antidepressants which increase neurotransmitters often increase neuroplasticity, too, so this may be actually why they work as they do, long after transmitter levels are raised.

It's not like refuelling a car, it's more like fertilising plants; it takes time for the helpful elements to be absorbed into the system.

All of these theories may contribute to, or may be consequences rather than causes of, depression. Research is ongoing. What is clear is that it's a very real, often extremely debilitating condition. Aside from cripplingly awful moods, depression also impairs cognitive ability. Many medical practitioners are taught how to differentiate between depression and dementia, as on cognitive tests serious memory problems and being genuinely unable to muster up any motivation to complete a test look the same, as far as the results are concerned. It's important to differentiate; the treatment for depression and dementia vary considerably, although often a diagnosis of dementia *leads to* depression,[8] which just complicates matters further.

Other tests show that people with depression pay more attention to negative stimuli.[9] If shown a list of words, they'll focus far more on those with unpleasant meanings ('murder', for example) than neutral ones ('grass'). We've discussed the brain's egocentric bias, meaning we focus on things that make us feel good about ourselves and ignore things that don't. Depression flips this: anything positive is ignored or downplayed; anything negative is perceived as 100 per cent accurate. As a result, once depression occurs, it can be extremely hard to get rid of.

While some people do seem to develop depression 'out of the blue', for many it's a consequence of too much time being hammered by life. Depression often occurs in conjunction with other serious conditions, including cancer, dementia and paralysis. There's also the famous 'downward spiral', where people's problems mount up over time. Losing your job is unpleasant, but if then your partner leaves you soon after, then a relative dies and you get mugged while heading home from the funeral, this can be just too much to deal with. The comfortable biases and assumptions our brains maintain to keep us motivated (that the world is fair, that nothing bad will happen to us) are shattered. We've no control over events, which makes matters worse. We stop seeing friends and pursuing interests, maybe turn to alcohol and drugs. All this, despite providing fleeting relief, taxes the brain further. The spiral continues.

These are risk factors for depression, which increase the likelihood of it occurring. Having a successful and public lifestyle, where money is no object and millions admire you, will have fewer risk factors than living in a deprived high-crime area, earning barely enough to survive and with no family support. If depression were like lightning, some people are indoors while others are struck outside near trees and flagpoles; the latter are more likely to get struck.

A successful lifestyle doesn't provide immunity. If someone rich and famous admits they suffer from depression, saying, 'How can they be depressed? They've got everything going for them', makes no sense. Being a smoker means you're more *likely* to develop lung cancer, but it doesn't affect *only* smokers. The brain's complexity mean many risk factors for depression aren't linked to your situation. Some have personality traits (such as a tendency to be self-critical) or even genes (depression is known to have a heritable component[10]) that make depression more likely.

What if the constant struggle against depression is what spurred someone to be successful? Staving off and/or overcoming depression often requires considerable willpower and effort, which can be channelled in interesting directions. The 'tears of a clown' cliché about successful comics whose skills stem from fighting internal torment is a perfect example, as are many famous creatives who endured the condition (Van Gogh, for instance). Far from a preventative, success may *result* from depression.

Also, unless you're born to it, achieving wealth and fame is hard work. Who knows what sacrifices a person made to obtain their success? And what if they eventually realise it wasn't worth it? Achieving something you've worked for for years can rip the purpose and drive from your life, leaving you adrift. Or, if you've lost the people you value on your determined upward career path, this can eventually be seen as too high a price. Being successful in other people's eyes is no defence. A healthy bank balance does not overrule the processes underlying depression. Even if it did, where's the cut-off point? Who would be 'too successful' to be ill? If you can't be depressed because you're better off than others, logically only the most unfortunate person on earth should be depressed.

This isn't to say many rich and successful people aren't very happy; it's just not a guarantee. The workings of your brain don't drastically change because you have a film career.

Depression is *not* logical. Those describing suicide and depression as selfish apparently struggle with this concept, as if those with depression make a table or chart with the pros and cons of suicide and, despite there being more cons, selfishly opt for suicide anyway.

This is nonsensical. A big problem with depression, perhaps *the* problem, is that it prevents you from behaving or thinking 'normally'. A person with depression is not thinking like a non-sufferer, in the same way that someone who's drowning is not 'breathing air' like someone on land. Everything we perceive and experience is processed and filtered through our brain, and if our brain has determined that everything is absolutely awful, that'll impact on everything else in our lives. From a depressed person's perspective, their self-worth may be so low, their outlook so bleak, that they genuinely believe their families/friends/fans would be better off without them in the world, that their suicide is actually an act of generosity. It's a very upsetting conclusion, but not one arrived at by a mind that's thinking 'straight'.

Accusations of selfishness also often imply people with depression are somehow choosing their situation, that they could enjoy life and be happy but consider it more convenient not to? Exactly how or why they'd do this is rarely explained. In instances of suicide, you get people saying it's the 'easy way out'. There are many ways to describe the sort of suffering that overrides millions-of-years-old survival instincts, but 'easy' isn't an obvious one. Perhaps none of it makes sense from a logical perspective, but insisting on logical thinking from someone in the grip of mental illness is like insisting that someone with a broken leg walks normally.

Depression isn't visible or communicable like a typical illness, so it's easier to deny that it's a problem rather than to accept the harsh unpredictable reality. Denial reassures the observer that 'it will never happen to me', but depression is still affects millions of people regardless, and throwing accusations of selfishness or laziness at them purely to make yourself feel better can't help. As a behaviour, that's a much better example of selfishness.

Sadly, the truth is that many persist in thinking it's easy to ignore or override a powerful debilitating mood disorder that regularly affects sufferers to the very core of their being. It's an excellent demonstration of how the brain values consistency, that once a person is decided on a certain viewpoint, it's hard to alter it. The people demanding that those with depression alter their thinking while refusing to do the same in the face of the evidence are showing just how hard it is. It's a terrible shame that those who are suffering the most get made to feel worse because of it.

It's bad enough when you've got your own brain conspiring against you so severely. To have other people's doing it too, that's just obscene.

Emergency shutdown
(Nervous breakdowns, and how they come about)

If you go out in the cold weather without a coat, you'll catch a cold. Junk food will mess up your heart. Smoking ruins your lungs. A poorly set-up workstation causes carpal tunnel and backache. Always lift with your knees. Don't crack your knuckles or you'll get arthritis. And so on.

You've probably heard these things before, and countless similar nuggets of wisdom about how to stay healthy. While the accuracy of these claims varies considerably, the notion that our actions affect our health is valid. Our bodies, marvellous as they are, have physical and biological limitations, and pushing these limits has consequences. So we watch what we eat, where we go, how we behave. If our bodies can be badly affected by what we do, what's preventing the same from happening to our complex, delicate brains? The answer is, of course, nothing.

In the modern world, the biggest threat to the well-being of our brains is good old stress.

Everyone experiences stress on a regular basis, but if it's too intense or too frequent, then we get problems. Chapter 1 explained how stress has very real and tangible effects on our health. Stress activates the hypothalamic–pituitary–adrenal (HPA) axis in the brain which activates fight-or-flight responses, which release adrenalin and cortisol, the 'stress' hormone. These have numerous effects on brain and body, so the effects of constant stress become very apparent in people. They're tense, not thinking straight, volatile, physically wasted or exhausted, and more. Such people are often said to be 'heading for a nervous breakdown'.

'Nervous breakdown' isn't an official medical or psychiatric term. It doesn't involve the literal breakdown of nerves. Some use 'mental breakdown', which is technically more accurate, but still a colloquialism. Regardless, most people will understand what it is. A nervous breakdown is what happens when someone can no longer cope with a high-stress situation, and just . . . 'snaps'. They 'shut down', 'withdraw', 'fall apart', 'can't cope'. It means a person is mentally no longer able to function as normal.

The experience of a nervous breakdown varies considerably between individuals. Some experience bleak depression, others crippling anxiety and panic attacks, some even hallucinations and psychosis. So it might be surprising that some see nervous breakdowns as a defence mechanism of the brain. As unpleasant as they are, they're potentially helpful.

Physiotherapy can be exhausting, hard and unpleasant, but it's certainly much better than not doing it. Nervous breakdowns may be the same, and this makes more sense when you consider that nervous breakdowns are invariably caused by stress.

We know how the brain experiences stress, but how does something cause stress in the first place? In psychology, things that cause stress are known (logically) as stressors. A stressor reduces personal control. Feeling in control makes most people feel secure and safe. It doesn't matter how much *actual* control we have. Every human is technically a meaningless sack of carbon clinging to a rock hurtling through the uncaring void around trillions of tonnes of nuclear fire, but that's too big for a single human to be aware of. But if we can demand and get soy milk in our latte, that's tangible control.

Stressors reduce options for action; something is more stressful if there's nothing you can do about it. Getting rained on is irksome if you have an umbrella. Getting caught in the rain without an umbrella while locked out of your house? That's stressful. With a headache or a cold there are medications available to minimise the symptoms, but chronic illnesses cause a lot of stress because there's often nothing to be done about them. They're a constant source of unavoidable unpleasantness, providing a very stressful situation.

A stressor also causes fatigue. Whether frantically running to catch a train after oversleeping or working on an important last-minute assignment, dealing with a stressor (and its physical consequences) requires energy and effort, depleting your reserves, causing further stress.

Unpredictability is also stressful. For example, epilepsy can cause incapacitating seizures at any moment, so they're impossible to plan for effectively, which is a stressful situation. It doesn't have to be a medical condition; living with a partner prone to mood swings or irrational behaviour, meaning you risk a rage-induced row with someone you love if you accidentally put the coffee jar in the wrong cupboard, can be incredibly stressful. These situations provide unpredictability and uncertainty, so we end up constantly on edge, expecting the worst at any moment. Result: stress.

Not all stress is debilitating. Most stress is manageable as we have compensatory mechanisms to balance the stress reactions. Cortisol stops being released; the parasympathetic nervous system activates to relax us again; we replenish our energy reserves; then carry on with our lives. However, in our complicated, interconnected modern world, there are many ways in which stress can quickly become overwhelming.

In 1967, Thomas Holmes and Richard Rahe assessed thousands of medical patients and asked them about their life experiences, attempting to establish a link between stress and illness.[11] They succeeded. This data lead to the formation of the Holmes and Rahe Stress Scale, where certain events are assigned a certain number of 'life change units' (LCU). The more LCU an event has, the more stressful it is. A person then says how many of the events on the scale happened to them in the previous year, and an overall score is assigned. The higher the score, the more likely someone is to become ill from stress. Top of the list is 'death of a spouse' with 100 LCU. Personal injury scores 53, getting sacked 47, trouble with in-laws 29, and so on. Surprisingly, divorce scores 73, whereas imprisonment scores 63. Oddly romantic, in a way.

Things not on the list can be worse again. A car crash, involvement in a violent crime, experiencing a major tragedy – these can cause 'acute' stress, where a single incident causes intolerable levels of stress. The events are so unexpected and traumatising that the usual stress response is, to quote *Spinal Tap*, 'turned up to 11'. The physical consequences of the fight-or-flight response are maximised (you often see someone shaking uncontrollably after serious trauma), but it's the effect on the brain that makes such extreme stress hard to get over. The flood of cortisol and adrenalin in the brain briefly enhances the memory system, producing 'flashbulb' memories. It's actually a useful evolved mechanism; when something severely stress-inducing happens, we definitely don't want to experience it again, so the highly stressed brain encodes as vivid and detailed a memory of it as possible, so we won't forget and blunder into it again. Makes sense, but in extremely stressful experiences it backfires; the memory's so vivid, and *remains* so vivid, that the individual keeps re-experiencing it, as if it were constantly reoccurring.

You know when you look at something extremely bright and it lingers in your vision because it was so intense it's 'burned' onto your retinas? This is the memory equivalent of that. Except it doesn't fade, it persists, because it's a *memory*. That's the point, and the memory is almost as traumatic as the original incident. The brain's system for preventing reoccurrence of trauma causes reoccurrence of trauma.

The constant stress caused by vivid flashbacks often results in numbing or dissociation, where people become detached from others, from experiencing emotions, even from reality itself. This is seen as another brain defence mechanism. Life is too stressful? Fine, shut it out, go into 'standby'. While effective in the short term, it's not a good long-term strategy. It impairs all manner of cognitive and behavioural faculties.

Post-Traumatic Stress Disorder (PTSD) is the most well-known consequence of this occurrence.[12]

Thankfully, most people won't experience such major traumas. Consequently, stress has to be sneakier to incapacitate them. So there's chronic stress, which is where you get one or more stressors that are more persistent than traumatic, so they affect you over the long term. A sick family member to care for, a tyrannical boss, a never-ending stream of deadlines, living on the breadline and never clearing your debts, these are all chronic stressors.*

This is bad, because when too much stress occurs over a long period, your ability to compensate suffers. The fight-or-flight mechanism actually becomes a problem. After a stressful event it typically takes the body 20–60 minutes to return to normal levels, so stress is quite long-lasting as it is.[14] The parasympathetic nervous system, which counteracts the fight-or-flight response once it's no longer needed, has to work hard to undo the effects of stress. When chronic stressors keep pumping stress hormones into our system, the parasympathetic nervous system is exhausted, so the physical and mental consequences of stress become 'normal'. Stress

* Most people experience stress via the workplace, which is odd. Stressing your employees should be a terrible thing for productivity. However, stress and pressure actually *increases* performance and motivation. Many people say they work better with a deadline, or do their best work under pressure. This isn't just an idle boast: in 1908, psychologists Yerkes and Dodson discovered that stressful situations actually *increase* performance on a task.[13] Consequences to avoid, fear of punishment, among other things,provide motivation and focus, improving someone's ability to do the job.

But only up to a point. Beyond that, when stress is too much, performance declines, and the more stress, the more it declines. This is known as the Yerkes–Dodson law. Many employers seem to understand the Yerkes–Dodson law intuitively, except for the 'too much stress makes things worse' part. It's like salt: some can improve your food, but too much overwhelms everything, ruining texture, taste and health.

hormones are no longer regulated and used when needed; they persist, and the person becomes constantly sensitised, twitchy, tense and distractible as a result.

The fact that we can't counteract stress internally means we seek external relief. Sadly, but predictably, this often makes things worse. This is known as the 'stress cycle', where attempts to alleviate stress actually cause more stress and consequences, which result in more attempts to reduce stress, which in turn cause more problems, and so on.

Say you get a new boss who assigns you more work than is reasonable. This would cause stress. But said boss is not open to reason or rational argument, so you work longer hours. You spend more time working and stressed, so you experience chronic stress. Soon you start consuming more junk food and alcohol to unwind. This negatively affects your health and mental state (junk food makes you unfit, alcohol is a depressant), which stresses you out further and makes you vulnerable to further stressors. So you get more stressed, and the cycle continues.

There are numerous ways to stop the ever-increasing stress (adjusting workloads, improved healthy lifestyle, therapeutic assistance, among others), but for many this just doesn't happen. So everything builds up, until a threshold is crossed and the brain essentially surrenders; much like a circuit breaker will cut the power before a surge overloads the system, so ever-increasing stress (with associated health consequences) would be terribly damaging for brain and body, so the brain puts a stop to, essentially, everything. Many argue the brain induces a nervous breakdown to stop stress escalating to the point where lasting damage can occur.

The threshold between 'stressed' and '*too* stressed' is hard to specify. There's the diathesis-stress model, where diathesis means 'vulnerability', which describes how someone who is more vulnerable to stress requires less stress to push them over the edge, into a full breakdown where they experience a mental disorder or 'episode' of some description. Some people are more susceptible: those with more difficult situations or lives; those already prone to paranoia or anxiety; even those with tremendous self-confidence can be brought low very quickly (if you're very self-confident, losing control due to stress could undermine your whole sense of self, causing immense stress).

Exactly how a nervous breakdown plays out also varies. Some people have an underlying condition like (or predisposition to) depression or anxiety, and overly stressful events can bring this on. Dropping a textbook on your toe hurts; dropping it on an already-fractured toe hurts considerably more. For some, the stress causes their mood to plummet to a point where it's incapacitating, and thus depression occurs. For others, the constant apprehension and persistence of stressful occurrences causes crippling anxiety or panic attacks. The cortisol released by stress is also known to have an effect on the dopamine systems of the brain,[15] making them more active and more sensitive. Anomalous activity in the dopamine systems are believed to be the underlying cause of psychosis and hallucinations, and some nervous breakdowns do produce psychotic episodes.

Thankfully, a nervous breakdown is typically a short-lived thing. Medical or therapeutic intervention usually sees people return to normal eventually, or just the enforced break from stress may help. Granted, not everyone sees a nervous breakdown as a helpful thing; not everyone gets over it, and those who do often retain a sensitivity to stress and adversity that means they could more easily experience a nervous breakdown again.[16] But they can at least resume a normal life, or a close approximation of it. Hence nervous breakdowns can help prevent lasting damage from a relentlessly stress-filled world.

Saying that, much of the problems a nervous breakdown helps limit are themselves caused by the brain's own techniques for dealing with stress, which often aren't up to scratch for modern life. Appreciating the brain for limiting the damage caused by stress via nervous breakdowns is like thanking someone for helping put out the fire in your house when they were the one who left the fryer on.

Dealing with the monkey on your back
(How the brain brings about drug addiction)

In the US in 1987, there was a televised public-service announcement that illustrated the dangers of drugs via the use of, surprisingly, eggs. An egg was shown and the viewer was told, 'This is your brain.' A frying pan was then shown, with the sentence, 'This is drugs.' The egg was then fried in the pan, with the words, 'This is your brain on drugs.' In a publicity sense, it was very successful. It won awards, and is still referenced (and, admittedly, mocked) in pop culture to this day. In a neuroscientific sense, it was a terrible campaign.

Drugs do not heat your brain so much that the very proteins making up its structure break down. Also, it's very rare for a drug to affect every part of the brain simultaneously, in the way that a frying pan affects an egg. Lastly, you apply drugs to the brain without removing it from its shell, aka skull. If it weren't, drug use certainly wouldn't so popular.

This isn't to say drugs are necessarily good for the brain; it's just the truth is far more complicated than egg-based metaphors can allow for.

The illegal drug trade is estimated at nearly half a *trillion* dollars[17] and many governments spend countless millions finding, destroying, and discouraging the use of illegal drugs. Drugs are widely assumed to be dangerous; they corrupt users, damage health and ruin lives. This is fair because drugs often do exactly that. Because they *work*. They work very well, and do so by altering and/or manipulating the fundamental processes of our brains. This causes problems such as addiction, dependence, behavioural changes and more, all of which stem from how our brains deal with drugs.

In Chapter 3, the dopaminergic mesolimbic pathway was mentioned. It's often called the 'reward' pathway or similar, because its function is refreshingly clear: it rewards us for actions perceived as being positive, by causing the sensation of pleasure. If we ever experience something enjoyable, from a particularly pleasant satsuma to the climax of a certain bedroom-based activity, the reward pathway provides the sensations that make us think, 'Well, wasn't that pleasant?'

The reward pathway can be activated by things we consume. Nutrition, hydration, alleviating appetite, providing energy; edible substances that do these things are recognised as pleasant because their beneficial actions trigger the reward pathway. For example, sugars provide easily utilised energy for our bodies, so sweet-tasting things are perceived as pleasant. The current state of the individual also plays a part: a glass of water and slice of bread would usually be considered the most uninspiring meal, but would be divine ambrosia to someone just washed up after months adrift at sea.

Most of these things activate the reward pathway 'indirectly', by causing a reaction in the body that the brain recognises as a good thing, thus warranting a rewarding sensation. Where drugs have the advantage, and what makes them dangerous, is they can activate the reward pathway 'directly'. The whole tedious process of 'having some positive effect on the body that the brain recognises' is skipped, like a bank employee handing over bags of cash without needing boring details like 'account numbers' or 'ID'. How does this happen?

Chapter 2 discussed how neurons communicate with each other via specific neurotransmitters, including noradrenaline, acetylcholine, dopamine, serotonin. Their job is to pass signals between neurons in a circuit or network. Neurons squirt them into synapses (the dedicated 'gap' between neurons where communication between them occurs). There they interact with dedicated receptors like a specific key opening a specific lock. The nature and type of receptor the transmitter interacts with determines the activity that results. It could be an excitatory neuron, which activate other regions of the brain like someone flicking a light switch, or it could be an inhibitory neuron, which reduces or shuts down activity in associated areas.

But suppose those receptors weren't as 'faithful' to specific neurotransmitters as hoped. What if other chemicals could mimic neurotransmitters, activating specific receptors in their absence? If this were possible, we could feasibly use these chemicals to manipulate the activity of our brains artificially. Turns out, it is possible, and we do it regularly.

Countless medications are chemicals that interact with certain cell receptors. Agonists cause receptors to activate and induce activity; for example, medications for slow or irregular heartbeats often involve substances that mimic adrenalin, which regulates cardiac activity. Antagonists occupy receptors but don't induce any activity, 'blocking' them and preventing genuine neurotransmitters from activating them, like a suitcase wedged in a lift door. Antipsychotic medications typically work by blocking certain dopamine receptors, as abnormal dopamine activity is linked to psychotic symptoms.

What if chemicals could 'artificially' induce activity in the reward pathway, without us having to do anything? They'd probably be very popular. So popular, in fact, that people would go to extreme lengths to get them. This is exactly what most drugs of abuse do.

Given the incredible diversity of beneficial things that we can do, the reward pathway has an incredibly wide variety of connections and receptors, meaning it's susceptible to a similarly wide variety of substances. Cocaine, heroin, nicotine, amphetamines, even alcohol – these all increase activity in the reward pathway, inducing unwarranted but undeniable pleasure. The reward pathway itself uses dopamine for all its functions and processes. As a result, numerous studies have shown that drugs of abuse invariably produce an increase in dopamine transmission in the reward pathway. This is what makes them 'enjoyable' – particularly drugs that mimic dopamine (cocaine, for example).[18]

Our powerful brains give us the intellectual capacity to quickly figure out that something induces pleasure, quickly decide we want more of it, and quickly work out how to get it. Luckily, we also have higher-brain regions in place to mitigate or overrule such base impulses as, 'Thing make me feel nice, must get more thing.' These impulse-control centres aren't perfectly understood but are most probably located in the prefrontal cortex, along with other complex cognitive functions.[19] Regardless, impulse control allows us to curb our excesses and recognise that descending into pure hedonism is not a good idea overall.

Another factor here is the plasticity and adaptability of the brain. A drug causes excess activity of a certain receptor? The brain responds by suppressing the activity of the cells those receptors activate, or shutting down the receptors, or doubling the number of receptors required to trigger a response, or any method that means 'normal' levels of activity are resumed. These processes are automatic; they don't differentiate between drug and neurotransmitter.

Think of it like a city hosting a major concert. Everything in the city is set up to maintain normal activity. Suddenly, thousands of excitable people arrive, and activity quickly becomes chaotic. In response, officials increase police and security presence, close roads, buses become more frequent, bars open earlier and close later, and so on. The excitable concert-goers are the drug, the brain is the city; too much activity and the defences kick in. This is 'tolerance', where the brain adapts to the drug so it no longer has the same potent effect.

The problem is, increased activity (in the reward pathway) is the whole point of a drug, and if the brain adapts to prevent this, there's only one solution: *more* drug. An increased dose is needed to provide the same sensation? Then that's what you use. Then the brain adapts to that, so you need a bigger dose. Then the brain adapts to that, and on it goes. Soon, your brain and body are so tolerant of a drug that you're taking doses that would legitimately kill someone who had never tried it before, but all it does is provide the same buzz that got you hooked in the first place.

This is one reason why quitting a drug, 'going cold turkey', is so challenging. If you're a long-term drug user, it's not a simple matter of willpower and discipline; your body and brain are now so used to the drug they've *physically altered to accommodate it*. Sudden removal of the drug therefore has serious consequences. Heroin and other opiates provide a good example of this.

Opiates are powerful analgesics that suppress normal levels of pain by stimulating the brain's endorphin (natural painkilling, pleasure-inducing neurotransmitters) and pain-management systems, providing an intense euphoria. Unfortunately, pain exists for a reason (to let us know about harm or damage), so the brain responds by increasing the potency of our pain-detection system, to cut through the blissful cloud of opiate-induced pleasure. So users take more opiates to shut it down again, and the brain strengthens it further, and so on.

Then the drug is taken away. The user no longer has something that made them incredibly calm and relaxed. What they do have is *a super-enhanced pain detection system*! Their pain-system activity is strong enough to cut through an opiate high, which for a normal brain would be agonising, as it is for a drug user going through withdrawal. Other systems affected by the drug are similarly altered. This is why cold turkey is so hard, and legitimately dangerous.

It would be bad enough if it was just these physiological changes that drugs cause. Alas, changes in the brain also alter behaviour. You'd think the many unpleasant consequences and demands of drug use should logically be sufficient to stop people using them. However, 'logic' is one of the first casualties of drug use. Parts of the brain may work to build tolerance and maintain normal functioning, but it's so diverse that other brain areas are simultaneously working to ensure we keep taking the drug. For example, it can cause the opposite of tolerance; drug users become sensitised to the effects of a drug by suppression of the adaptation systems,[20] so it becomes *more* potent, compelling the individual to seek it out even more. This is one factor that leads to addiction.*

There's more. Communication between the reward pathway and the amygdala serves to provide a strong emotional response to anything drug related, aka 'drug cues'.[22] Your specific pipe, syringe, lighter, the smell of the substance, all these become emotionally charged and stimulating in their own right. This means drug users can experience the effects of a drug, directly from the things *associated* with it.

Heroin addicts provide another grim example of this. One treatment for heroin addiction is methadone, another opiate that provides similar (though reduced) effects, theoretically enabling users to give up gradually without going cold turkey. Methadone is supplied in a form than can only be swallowed (it looks like worryingly green cough syrup), whereas heroin is usually injected. But so strong a connection does the brain make between injection and the effects of heroin, that the act

* To clarify, you can be addicted to things other than drugs. Shopping, video games, anything that can activate the reward pathway above normal levels. Gambling addiction is a particularly bad one. Obtaining lots of money for minimal effort is very rewarding, but it's really hard to undo this addiction. Usually, it would involve long periods of no reward so the brain stops expecting it, but, with gambling, long periods of not winning is *normal*, as is losing money.[21] Consequently, it's hard to convince gambling addicts that gambling is bad, as they're already fully aware of this.

of injecting causes a high. Addicts have been known to pretend to swallow methadone, then spit it into a syringe and inject it.[23] This is an incredibly dangerous act (if only for hygiene reasons) but the warping of the brain by drugs means the method of delivery is almost as important as the drug itself.

Constant stimulation of the reward pathway by drugs also alters our ability to think and behave rationally. The interface between the reward pathway and the frontal cortex, where the important conscious decisions are made, is modified, so that drug-acquiring behaviours are prioritised above normally more important things (such as holding down a job, obeying the law, showering). By contrast, negative consequences of drugs (being arrested, getting a nasty illness from needle sharing, alienating friends and family) are actually suppressed in terms of how much they bother or worry us. Hence an addict will shrug nonchalantly at losing all their worldly possessions but will repeatedly risk their own skin to obtain another hit.

Perhaps most disconcerting is the fact that excessive drug use suppresses activity of the prefrontal cortex and impulse-control areas. The parts of the brain that say, 'Don't do that', 'That's not clever', 'You'll regret this', and so on – their influence is diminished. Free will may be one of the most profound achievements of the human brain, but if it gets in the way of a buzz then it's got to go.[24]

The bad news keeps coming. These drug-based alterations to the brain and all the associations made don't go away when drug use stops; they're just 'not used'. They may fade somewhat but they endure, and will still be there should the individual sample the drug again, no matter how long they've abstained. This is why relapse is so easy, and such a big problem.

Exactly how people end up becoming regular drug takers varies massively. Maybe they live in bleak deprived areas where the only relief from the realities of life is from drugs. They might have an undiagnosed mental disorder and end up 'self-medicating' by trying drugs to alleviate the problems they experience every day. There is even believed to be a genetic component to drug use, possibly due to some people having a less-developed or underpowered impulse-control region of the brain.[25] Everyone has that part of them that, when offered the opportunity to try a new experience, says, 'What's the worst that could happen?' Sadly, some people lack that other part of the brain that explains in exquisite detail exactly what could happen. This accounts for why many people can safely dabble with drugs and walk away unchanged, while others are ensnared from the first hit onwards.

Regardless of the cause or initial decisions that lead to it, addiction is recognised by professionals as a condition to be treated rather than a failing to be criticised or condemned. Excessive drug use causes the brain to undergo startling changes, many of which contradict each other. Drugs seem to turn the brain against itself in some prolonged war of attrition, where our lives are the battleground. This is a terrible thing to do to yourself, but drugs make it so that you don't care.

This is your brain on drugs. It is pretty hard to convey all this with eggs, admittedly.

Reality is overrated anyway (Hallucinations, delusions and what the brain does to cause them)

One of the most common occurrences in mental health problems is psychosis, where someone's ability to tell what's real or not is compromised. The most common expressions of this are hallucinations (perceiving something that isn't actually there) and delusions (unquestionably believing something that is demonstrably not true), along with other behavioural and thought disruptions. The idea of these things happening can be deeply unsettling; losing your very grasp on reality itself, how are you supposed to deal with that?

Worryingly, the neurological systems handling something as integral as the ability to grasp reality are disturbingly vulnerable. Everything covered in this chapter so far – depression, drugs and alcohol, stress and nervous breakdowns – can end up triggering hallucinations and delusions in the overtaxed brain. There are also many other things that trigger them, like dementia, Parkinson's disease, bipolar disorder, lack of sleep, brain tumours, HIV, syphilis, Lyme disease, multiple sclerosis, abnormally low blood sugar alcohol, cannabis, amphetamines, ketamine, cocaine, and more. Some conditions are so synonymous with psychosis they're known as 'psychotic disorders', the most well known of which is schizophrenia. To clarify, schizophrenia isn't about split personalities; the 'schism' for which it is named is more between the individual and reality.

While psychosis often results in the sensation of being touched when you're not being, or tasting or smelling things that aren't there, the most common are aural hallucinations, aka 'hearing voices'. There are several classes of this type of hallucination.

There are first-person auditory hallucinations ('hearing' your own thoughts, as if they're spoken by someone else), second person (hearing a separate voice talking *to* you) and third person (hearing one or more voices talking *about* you, providing a running commentary ofnwhat you're doing). The voices can be male or female, familiar or unfamiliar, friendly or critical. If the latter is the case (which it usually is), they are 'derogatory' hallucinations. The nature of hallucinations can help diagnosis; for instance, persistent derogatory third-person hallucinations are a reliable indicator of schizophrenia.[26]

How does this happen? It's tricky to study hallucinations, because you'd need people to hallucinate on cue in the lab. Hallucinations are generally unpredictable, and if someone could switch them on and off at will, they wouldn't be a problem. Nevertheless, there have been numerous studies, focusing largely on the auditory hallucinations experienced by those with schizophrenia, which tend to be very persistent.

The most common theory of how hallucinations occur focuses on the complex processes the brain uses to differentiate between neurological activity generated by the outside world, and activity we generate internally. Our brains are always chattering away, thinking, musing, worrying and so on. This all produces (or is produced by) activity within the brain.

The brain is usually quite capable of separating internal from external activity (that produced by sensory information), like keeping received and sent emails in separate folders. The theory is that hallucinations occur when this ability is compromised. If you've ever accidentally lumped all your emails together in the same folder you'll know how confusing this can be, so imagine doing that with your brain functions.

So the brain loses track of what's internal and what's external activity, and the brain isn't good with such things. This was demonstrated in Chapter 5, which discussed how blindfolded people struggle to tell the difference between apples and potatoes when eating them. That's the brain functioning 'normally'. In the case of hallucinations, the systems that separate internal and external activity are (metaphorically) blindfolded. So people end up perceiving internal monologue as an actual person speaking, as internal musings and hearing spoken words activates the auditory cortex and associated language-processing areas. Indeed, a number of studies have shown that persistent third-person hallucinations correspond with reduced volumes of grey matter in these areas.[27] Grey matter does all the processing, so this suggests reduced ability to distinguish between internally and externally generated activity.

Evidence for this comes from an unlikely source: tickling. Most people can't tickle themselves. Why not? Tickling should feel the same no matter who does it, but tickling ourselves involves conscious choice and action on our part, which requires neurological activity, which the brain recognises as being internally generated, so it's processed differently. The brain detects the tickling, but internal conscious activity flagged it up beforehand, so it's ignored. As such, it provides a useful example of the brain's ability to differentiate between internal and external activity. Professor Sarah-Jayne Blakemore and her colleagues at the Wellcome Department of Cognitive Neurology studied the ability of psychiatric patients tickle themselves.[28] They found that, compared with non-patients, patients who experienced hallucinations were far more sensitive to self-tickling, suggesting a compromised ability to separate internal and external stimuli.

While an interesting approach (and one not without flaws), please note that being able to tickle yourself does not automatically mean you're psychotic. People vary tremendously. My wife's university housemate could tickle himself, and has never had any psychiatric issues. He's extremely tall though; maybe the nerve signals take so long to get to the brain from the tickling site it just forgets how they originated?*

Neuroimaging studies have suggested further theories about how hallucinations generally come about. An extensive review of the available evidence, published by Dr Paul Allen and his colleagues in 2008,[29] suggests an intricate (but surprisingly logical) mechanism.

As you may expect, our brain's ability to differentiate between internal and external occurrences is derived from multiple areas acting together. There are fundamental subcortical areas, predominantly the thalamus, that provide raw information from the senses. This ends up in the sensory cortex, which is an umbrella term for all the different areas involved in sensory processing (the occipital lobe for vision, auditory and olfactory processing in the temporal lobes, and so on). It's often subdivided into primary and secondary sensory cortex; primary processes the raw features of a stimulus, secondary processes more fine detail and recognition (for example, the primary sensory cortex would recognise specific lines, edges and colours, the secondary would recognise all of this as an oncoming bus, so both are important).

* This isn't remotely possible. I came up with this theory as a student when put on the spot. In those days, I was far more arrogant and would rather make ridiculous wild guesses than admit to not knowing something.

Connecting to the sensory cortex are areas of the prefrontal cortex (decisions and higher functions, thinking), premotor cortex (producing and overseeing conscious movement), cerebellum (fine motor control and maintenance) and regions with similar functions. These areas are generally responsible for determining our conscious actions, providing information needed to determine which activity is internally generated, as in the tickling example. The hippocampus and amygdala also incorporate memory and emotion, so we can remember what we're perceiving and react accordingly.

Activity between these interconnected regions maintains our ability to separate the outside world from the one inside our skull. It's when the connections are changed by something that affects the brain that hallucinations occur. Increased activity in the secondary sensory cortex means signals generated by internal processes get stronger and affect us more. Reduced activity from the connections to the prefrontal cortex, premotor cortex, and so on, prevents the brain from recognising information that is produced internally. These areas are also believed to be responsible for monitoring the external/internal detection system, ensuring genuine sensory information is processed as such, so compromised connections with these areas would mean more internally-generated information is 'perceived' as genuine.[30]

All of this combined causes hallucinations. If you think to yourself, 'That was stupid', when you buy an expensive new tea set and let your toddler carry it out of the shop, this is usually processed as an internal observation. But if your brain wasn't able to recognise that it came from the prefrontal cortex, the activity it produces in the language-processing areas could be recognised as something spoken. Atypical amygdala activity means the emotional associations of this wouldn't be dampened either, so we end up 'hearing' a very critical voice.

The sensory cortex processes everything and internal activity can relate to anything, so hallucinations occur in all senses. Our brains, knowing no better, incorporate all of this anomalous activity into the perception process so we end up perceiving alarming, unreal things that aren't there. With such a widespread network of systems responsible for our awareness of what's real and what isn't, it is undoubtedly vulnerable to a wide variety of factors, hence hallucinations in psychosis are so common.

Delusions, a false belief in something that is demonstrably untrue, are another common feature of psychosis, and again demonstrate a compromised ability to distinguish between real and not-real. Delusions have many forms, such as grandiose delusions, where an individual believes they're far more impressive than is accurate (believing they're a world-leading business genius despite being a part-time shoe-shop employee), or (more common) persecutory delusions, where an individual believes they're being relentlessly persecuted (everyone they meet is part of some shadowy plot to kidnap them).

Delusions can be as varied and strange as hallucinations, but are often far more stubborn; delusions tend to be 'fixed', and highly resistant to contradictory evidence. It's easier to convince someone the voices they're hearing aren't real than it is to convince a delusional person that not everyone is plotting against them. Rather than regulating internal and external activity, delusions are believed to stem from the brain's systems for interpreting what *does* happen and what *should* happen.

The brain has to deal with a lot of information at every given moment, and to do this effectively it maintains a mental model of how the world is meant to work. Beliefs, experiences, expectations, assumptions, calculations – all of these are combined into a constantly updated general understanding of how things happen, so we know what to expect and how to react without having to figure it out again each time. As a result, we're not constantly surprised by the world around us.

You walk along the street and a bus stops alongside you. This isn't surprising because your mental model of the world recognises and knows how buses operate; you know buses stop to let passengers on and off, so you ignore this occurrence. However, if a bus pulls up outside your house and doesn't move, this would be atypical. Your brain is now has new, unfamiliar information, and it needs to make sense of it in order to update and maintain the mental model of the world.

So you investigate, and it turns out the bus has broken down. But, before you discover this, a number of other theories will have occurred to you. The bus driver's spying on you? Someone bought you a bus? Your house has been designated as a bus depot without your knowledge? The brain comes up with all these explanations, but recognises them as very unlikely, based on the existing mental model of how things work, so they're dismissed.

Delusions result when this system undergoes alteration. A well-known type of delusion is Capgras delusion, where people genuinely believe someone close to them (spouse, parent, sibling, friend, pet) has been replaced by an identical impostor.[31] Usually when you see a loved one, this triggers multiple memories and emotions: love, affection, fondness, frustration, irritation (depending on length of relationship).

But suppose you see your partner and experience none of the usual emotional associations? Damage to areas of the frontal lobes can cause this to happen. Based on all your memories and experiences, your brain anticipates a strong emotional response to the sight of your partner, but this doesn't happen. This results in uncertainty: that's my long-term partner, I have many feelings about my long-term partner, feelings I'm now not experiencing. Why not? One way to resolve this inconsistency is the conclusion that they aren't your partner, but a physically identical impostor. This conclusion allows the brain to reconcile the disharmony it's experiencing, thus ending uncertainty. This is Capgras delusion.

The trouble is, it's clearly wrong, but the individual's brain doesn't recognise it as such. Objective proof of their partner's identity just makes the lack of emotional connection worse, so the conclusion that they're an impostor is even more 'reassuring'. Thus a delusion is sustained in the face of evidence.

This is the basic process believed to underlie delusions in general; the brain *expects* something to happen, it perceives something *different* happening, the expectations and occurrence don't match, a solution to this mismatch must be found. It starts to become problematic if solutions rely on ridiculous or unlikely conclusions.

Thanks to other stresses and factors upsetting the delicate systems of our brain, things we perceive that would usually be dismissed as innocuous or irrelevant end up being processed as far more significant. The delusions themselves can in fact suggest the nature of the problem producing them.[32] For example, excessive anxiety and paranoia would mean an individual is experiencing unexplained activation of the threat-detection and other defensive systems, so it would try to reconcile this by finding a source for the mysterious threat, and thus interpret harmless behaviour (for instance someone muttering to herself in a shop as you pass) as suspicious and threatening, provoking delusions of mysterious plots against them. Depression invokes inexplicable low mood, so any experiences that are even slightly negative (perhaps someone leaving a table just as you sit down next to them) become significant and are interpreted as people intense disliking you due to your awfulness, and thus delusions occur.

Things that don't conform to our mental model of how the world works are often downplayed or suppressed; they don't conform to our expectations or predictions, and the best explanation is that they're wrong, so can be ignored. You might believe there is no such thing as aliens, so anyone claiming to have seen UFOs or been abducted is dismissed as a raving idiot. Someone else's claims don't prove your beliefs are wrong. This works up to a point; should you then be abducted by aliens and vigorously probed, your conclusions are likely to change. But, in delusional states, the experiences that contradict your own conclusions can be even more suppressed than normal.

Current theories about the neurological systems responsible propose a frighteningly complex arrangement, stemming from another widespread network of brain areas (parietal lobe regions, prefrontal cortex, temporal gyrus, striatum, amygdala, cerebellum, mesocorticolimbic regions, and so on).[33] There's also evidence suggesting those prone to delusions show an excess of the excitatory (producing more activity) neurotransmitter glutamate, which may explain innocuous stimulation becoming overly significant.[34] Too much activity also exhausts neuronal resources, reducing neuronal plasticity, so the brain is less able to change and adapt the affected areas, making delusions more persistent again.

A word of caution: this section has focused on hallucinations and delusions being caused by disruptions and problems with the brain's processes, which does suggest that they're due only to disorders or illnesses. This isn't the case. You may think someone is 'deluded' if they believe the earth is only six thousand years old and dinosaurs never existed, but millions of people genuinely believe this. Similarly, some people genuinely believe their deceased relatives are talking to them. Are they sick? Grieving? Is this a coping mechanism? A spiritual thing? There are many possible explanations other than 'poor mental health'.

Our brains determine what's real or not based on our experiences, and if we grow up in a context where objectively impossible things are seen as normal, then our brains conclude they *are* normal, and judge everything else accordingly. Even people not brought up in the more extreme belief system are susceptible – the 'just world' bias described in Chapter 7 is incredibly common, and often leads to conclusions, beliefs and assumptions about people experiencing hardships that aren't correct.

This is why unrealistic beliefs are classed as delusions only if they're not consistent with the person's existing belief system and views. The experience of a devout evangelist from the American Bible Belt saying he can hear the voice of God is not considered a delusion. An agnostic trainee accountant from Sunderland saying she can hear the voice of God? Yes, she'll probably be classed as delusional.

The brain provides us with an impressive perception of reality but, as we've seen repeatedly throughout this book, much of this perception is based on calculations, extrapolations and sometimes outright guesswork on the brain's part. Given every possible thing that can affect how the brain does things, it's easy to see how such processes might go a bit awry, especially considering how what's 'normal' is more general consensus than fundamental fact. It's amazing how humans get anything done, really.

That's assuming they actually *do* get anything done. Maybe that's just what we tell ourselves for reassurance. Maybe nothing is real? Maybe this whole book has been a hallucination? All things being equal, I hope it isn't, or I've wasted quite a considerable amount of time and effort.

Afterword

So that's the brain. Impressive, isn't it? But, also, a bit stupid.

Acknowledgements

To my wife, Vanita, for supporting me in yet another ridiculous endeavour with only a bare minimum of eye-rolling.

To my children, Millen and Kavita, for giving me reason to want to try writing a book, and being too young to care if I succeed or not.

To my parents, without whom I wouldn't be able to do this. Or anything at all, when you think about it.

To Simon, for being a good enough friend to remind me this might end up being rubbish whenever I got too carried away with myself.

To my agent, Chris of Greene and Heaton, for all his hard work, and particularly getting in touch in the first place and saying, 'Have you ever thought about writing a book?', because I hadn't at that point.

To my editor, Laura, for all her efforts and patience, particularly for pointing out, 'You're a neuroscientist. You should write about the brain', repeatedly until I realised this made sense.

To John, Lisa and all the others at Guardian Faber for turning my ramshackle efforts into something people seemed to actually want to read.

To James, Tash, Celine, Chris and several more Jameses at the *Guardian*, for allowing me the opportunity to contribute to your major publication, despite my certainty that this was due to a clerical error.

To all other friends and family who offered support, help and essential distraction while I wrote this book.

You. All of you. This is technically all your fault.

References

1 Mind controls

1 S. B. Chapman et al., 'Shorter term aerobic exercise improves brain, cognition, and cardiovascular fitness in aging', *Frontiers in Aging Neuroscience*, 2013, vol. 5
2 V. Dietz, 'Spinal cord pattern generators for locomotion', *Clinical Neurophysiology*, 2003, 114(8), pp. 1379–89
3 S. M. Ebenholtz, M. M. Cohen and B. J. Linder, 'The possible role of nystagmus in motion sickness: A hypothesis', *Aviation, Space, and Environmental Medicine*, 1994, 65(11), pp. 1032–5
4 R. Wrangham, *Catching Fire: How Cooking Made Us Human*, Basic Books, 2009
5 'Two Shakes-a-Day Diet Plan – Lose weight and keep it off', http://www.nutritionexpress.com/article+index/diet+weight+loss/diet+plans+tips/showarticle.aspx?id=1904 (accessed September 2015)
6 M. Mosley, 'The second brain in our stomachs', http://www.bbc.co.uk/news/health-18779997 (accessed September 2015)
7 A. D. Milner and M. A. Goodale, *The Visual Brain in Action*, Oxford University Press, (Oxford Psychology Series no. 27), 1995
8 R. M. Weiler, 'Olfaction and taste', *Journal of Health Education*, 1999, 30(1), pp. 52–3
9 T. C. Adam and E. S. Epel, 'Stress, eating and the reward system', *Physiology & Behavior*, 2007, 91(4), pp. 449–58
10 S. Iwanir et al., 'The microarchitecture of C. elegans behavior during lethargus: Homeostatic bout dynamics, a typical body posture, and regulation by a central neuron', *Sleep*, 2013, 36(3), p. 385
11 A. Rechtschaffen et al., 'Physiological correlates of prolonged sleep deprivation in rats', *Science*, 1983, 221(4606), pp. 182–4
12 G. Tononi and C. Cirelli, 'Perchance to prune', *Scientific American*, 2013, 309(2), pp. 34–9
13 N. Gujar et al., 'Sleep deprivation amplifies reactivity of brain reward networks, biasing the appraisal of positive emotional experiences', *Journal of Neuroscience*, 2011, 31(12), pp. 4466–74
14 J. M. Siegel, 'Sleep viewed as a state of adaptive inactivity', *Nature Reviews Neuroscience*, 2009, 10(10), pp. 747–53

15 C. M. Worthman and M. K. Melby, 'Toward a comparative developmental ecology of human sleep', in M. A. Carskadon (ed.), *Adolescent Sleep Patterns*, Cambridge University Press, 2002, pp. 69–117

16 S. Daan, B. M. Barnes and A. M. Strijkstra, 'Warming up for sleep? – Ground squirrels sleep during arousals from hibernation', *Neuroscience Letters*, 1991, 128(2), pp. 265–8

17 J. Lipton and S. Kothare, 'Sleep and Its Disorders in Childhood', in A. E. Elzouki (ed.), *Textbook of Clinical Pediatrics*, Springer, 2012, pp. 3363–77

18 P. L. Brooks and J. H. Peever, 'Identification of the transmitter and receptor mechanisms responsible for REM sleep paralysis', *Journal of Neuroscience*, 2012, 32(29), pp. 9785–95

19 H. S. Driver and C. M. Shapiro, 'ABC of sleep disorders. Parasomnias', *British Medical Journal*, 1993, 306(6882), pp. 921–4

20 '5 Other Disastrous Accidents Related To Sleep Deprivation', http://www.huffingtonpost.com/2013/12/03/sleep-deprivation-accidents-disasters_n_4380349.html (accessed September 2015)

21 M. Steriade, *Thalamus*, Wiley Online Library, [1997], 2003

22 M. Davis, 'The role of the amygdala in fear and anxiety' *Annual Review of Neuroscience*, 1992, 15(1), pp. 353–75

23 A. S. Jansen et al., 'Central command neurons of the sympathetic nervous system: Basis of the fight-or-flight response', *Science*, 1995, 270(5236), pp. 644–6

24 J. P. Henry, 'Neuroendocrine patterns of emotional response', in R. Plutchik and H. Kellerman (eds), *Emotion: Theory, Research and Experience*, vol. 3: *Biological Foundations of Emotion*, Academic Press, 1986, pp. 37–60

25 F. E. R. Simons, X. Gu and K. J. Simons, 'Epinephrine absorption in adults: Intramuscular versus subcutaneous injection', *Journal of Allergy and Clinical Immunology*, 2001, 108(5), pp. 871–3

2 The gift of memory (keep the receipt)

1 N. Cowan, 'The magical mystery four: How is working memory capacity limited, and why?' *Current Directions in Psychological Science*, 2010, 19(1): pp. 51–7

2 J. S. Nicolis and I. Tsuda, 'Chaotic dynamics of information processing: The "magic number seven plus-minus two" revisited', *Bulletin of Mathematical Biology*, 1985, 47(3), pp. 343–65

3 P. Burtis, P., 'Capacity increase and chunking in the development of short-term memory', *Journal of Experimental Child Psychology*, 1982, 34(3), pp. 387–413

4 C. E. Curtis and M. D'Esposito, 'Persistent activity in the prefrontal cortex during working memory', *Trends in Cognitive Sciences*, 2003, 7(9), pp. 415–23
5 E. R. Kandel and C. Pittenger, 'The past, the future and the biology of memory storage', *Philosophical Transactions of the Royal Society of London B: Biological Sciences*, 1999, 354(1392), pp. 2027–52
6 D. R. Godden and A.D. Baddeley, 'Context-dependent memory in two natural environments: On land and underwater', *British Journal of Psychology*, 1975, 66(3), pp. 325–31
7 R. Blair, 'Facial expressions, their communicatory functions and neuro-cognitive substrates', *Philosophical Transactions of the Royal Society B: Biological Sciences*, 2003, 358(1431), pp. 561–72
8 R. N. Henson, 'Short-term memory for serial order: The start-end model', *Cognitive Psychology*, 1998, 36(2), pp. 73–137
9 W. Klimesch, *The Structure of Long-term Memory: A Connectivity Model of Semantic Processing*, Psychology Press, 2013
10 K. Okada, K. L. Vilberg and M. D. Rugg, 'Comparison of the neural correlates of retrieval success in tests of cued recall and recognition memory', *Human Brain Mapping*, 2012, 33(3), pp. 523–33
11 H. Eichenbaum, *The Cognitive Neuroscience of Memory: An Introduction*, Oxford University Press, 2011
12 E. E. Bouchery et al., 'Economic costs of excessive alcohol consumption in the US, 2006', *American Journal of Preventive Medicine*, 2011, 41(5), pp. 516–24
13 A. Ameer and R. R. Watson, 'The Psychological Synergistic Effects of Alcohol and Caffeine', in R. R. Watson et al., *Alcohol, Nutrition, and Health Consequences*, Springer, 2013, pp. 265–70
14 L. E. McGuigan, *Cognitive Effects of Alcohol Abuse: Awareness by Students and Practicing Speech-language Pathologists*, Wichita State University, 2013
15 T. R. McGee et al., 'Alcohol consumption by university students: Engagement in hazardous and delinquent behaviours and experiences of harm', in *The Stockholm Criminology Symposium 2012*, Swedish National Council for Crime Prevention, 2012
16 K. Poikolainen, K. Leppänen and E. Vuori, 'Alcohol sales and fatal alcohol poisonings: A time series analysis', *Addiction*, 2002, 97(8), pp. 1037–40
17 B. M. Jones and M. K. Jones, 'Alcohol and memory impairment in male and female social drinkers', in I. M. Bimbaum and E. S. Parker (eds) *Alcohol and Human Memory (PLE: Memory)*, 2014, 2, pp. 127–40
18 D. W. Goodwin, 'The alcoholic blackout and how to prevent it', in

I. M. Bimbaum and E. S. Parker (eds) *Alcohol and Human Memory*, 2014, 2, pp. 177–83

19 H. Weingartner and D. L. Murphy, 'State-dependent storage and retrieval of experience while intoxicated', in I. M. Bimbaum and E. S. Parker (eds) *Alcohol and Human Memory (PLE: Memory)*, 2014, 2, pp. 159–75

20 J. Longrigg, *Greek Rational Medicine: Philosophy and Medicine from Alcmaeon to the Alexandrians*, Routledge, 2013

21 A. G. Greenwald, 'The totalitarian ego: Fabrication and revision of personal history', *American Psychologist*, 1980, 35(7), p. 603

22 U. Neisser, 'John Dean's memory: A case study', *Cognition*, 1981, 9(1), pp. 1–22

23 M. Mather and M. K. Johnson, 'Choice-supportive source monitoring: Do our decisions seem better to us as we age?', *Psychology and Aging*, 2000, 15(4), p. 596

24 *Learning and Motivation*, 2004, 45, pp. 175–214

25 C. A. Meissner and J. C. Brigham, 'Thirty years of investigating the own-race bias in memory for faces: A meta-analytic review', *Psychology, Public Policy, and Law*, 2001, 7(1), p. 3

26 U. Hoffrage, R. Hertwig and G. Gigerenzer, 'Hindsight bias: A by-product of knowledge updating?', *Journal of Experimental Psychology: Learning, Memory, and Cognition*, 2000, 26(3), p. 566

27 W. R. Walker and J. J. Skowronski, 'The fading affect bias: But what the hell is it for?', *Applied Cognitive Psychology*, 2009, 23(8), pp. 1122–36

28 J. D⊠biec, D. E. Bush and J. E. LeDoux, 'Noradrenergic enhancement of reconsolidation in the amygdala impairs extinction of conditioned fear in rats – a possible mechanism for the persistence of traumatic memories in PTSD', *Depression and Anxiety*, 2011, 28(3), pp. 186–93

29 N. J. Roese and J. M. Olson, *What Might Have Been: The Social Psychology of Counterfactual Thinking*, Psychology Press, 2014

30 A. E. Wilson and M. Ross, 'From chump to champ: people's appraisals of their earlier and present selves', *Journal of Personality and Social Psychology*, 2001, 80(4), pp. 572–84

31 S. M. Kassin et al., 'On the "general acceptance" of eyewitness testimony research: A new survey of the experts', *American Psychologist*, 2001, 56(5), pp. 405–16

32 http://socialecology.uci.edu/faculty/eloftus/ (accessed September 2015)

33 E. F. Loftus, 'The price of bad memories', Committee for the Scientific Investigation of Claims of the Paranormal, 1998

34 C. A. Morgan et al., 'Misinformation can influence memory for recently experienced, highly stressful events', *International Journal*

of Law and Psychiatry, 2013, 36(1), pp. 11–17
35 B. P. Lucke-Wold et al., 'Linking traumatic brain injury to chronic traumatic encephalopathy: Identification of potential mechanisms leading to neurofibrillary tangle development', *Journal of Neurotrauma*, 2014, 31(13), pp. 1129–38
36 S. Blum et al., 'Memory after silent stroke: Hippocampus and infarcts both matter', *Neurology*, 2012, 78(1), pp. 38–46
37 R. Hoare, 'The role of diencephalic pathology in human memory disorder', *Brain*, 1990, 113, pp. 1695–706
38 L. R. Squire, 'The legacy of patient HM for neuroscience', *Neuron*, 2009, 61(1), pp. 6–9
39 M. C. Duff et al., 'Hippocampal amnesia disrupts creative thinking', *Hippocampus*, 2013, 23(12), pp. 1143–9
40 P. S. Hogenkamp et al., 'Expected satiation after repeated consumption of low- or high-energy-dense soup', *British Journal of Nutrition*, 2012, 108(01), pp. 182–90
41 K. S. Graham and J. R. Hodges, 'Differentiating the roles of the hippocampus complex and the neocortex in long-term memory storage: Evidence from the study of semantic dementia and Alzheimer's disease', *Neuropsychology*, 1997, 11(1), pp. 77–89
42 E. Day et al., 'Thiamine for Wernicke-Korsakoff Syndrome in people at risk from alcohol abuse', *Cochrane Database of Systemic Reviews*, 2004, vol. 1
43 L. Mastin, 'Korsakoff's Syndrome. The Human Memory – Disorders 2010', http://www.human-memory.net/disorders_korsakoffs.html (accessed September 2015)
44 P. Kennedy and A. Chaudhuri, 'Herpes simplex encephalitis', *Journal of Neurology, Neurosurgery & Psychiatry*, 2002, 73(3), pp. 237–8

3 Fear: nothing to be scared of

1 H. Green et al., *Mental Health of Children and Young People in Great Britain, 2004*, Palgrave Macmillan, 2005
2 'In the Face of Fear: How fear and anxiety affect our health and society, and what we can do about it, 2009', http://www.mentalhealth.org.uk/publications/in-the-face-of-fear/ (accessed September 2015)
3 D. Aaronovitch and J. Langton, *Voodoo Histories: The Role of the Conspiracy Theory in Shaping Modern History*, Wiley Online Library, 2010
4 S. Fyfe et al., 'Apophenia, theory of mind and schizotypy: Perceiving meaning and intentionality in randomness', *Cortex*, 2008, 44(10), pp. 1316–25

5 H. L. Leonard, 'Superstitions: Developmental and Cultural Perspective', in R. L. Rapoport (ed.), *Obsessive-compulsive Disorder in Children and Adolescents*, American Psychiatric Press, 1989, pp. 289–309

6 H. M. Lefcourt, *Locus of Control: Current Trends in Theory and Research (2nd edn)*, Psychology Press, 2014

7 J. C. Pruessner et al., 'Self-esteem, locus of control, hippocampal volume, and cortisol regulation in young and old adulthood', *Neuroimage*, 2005, 28(4), pp. 815–26

8 J. T. O'Brien et al., 'A longitudinal study of hippocampal volume, cortisol levels, and cognition in older depressed subjects', *American Journal of Psychiatry*, 2004, 161(11), pp. 2081–90

9 M. Lindeman et al., 'Is it just a brick wall or a sign from the universe? An fMRI study of supernatural believers and skeptics', *Social Cognitive and Affective Neuroscience*, 2012, pp.943–9

10 A. Hampshire et al., 'The role of the right inferior frontal gyrus: inhibition and attentional control', *Neuroimage*, 2010, 50(3), pp. 1313–19

11 J. Davidson, 'Contesting stigma and contested emotions: Personal experience and public perception of specific phobias', *Social Science & Medicine*, 2005, 61(10), pp. 2155–64

12 V. F. Castellucci and E. R. Kandel, 'A quantal analysis of the synaptic depression underlying habituation of the gill-withdrawal reflex in Aplysia', *Proceedings of the National Academy of Sciences*, 1974, 71(12), pp. 5004–8

13 S. Mineka and M. Cook, 'Social learning and the acquisition of snake fear in monkeys', *Social Learning: Psychological and Biological Perspectives*, 1988, pp. 51–73

14 K. M. Mallan, O. V. Lipp and B. Cochrane, 'Slithering snakes, angry men and out-group members: What and whom are we evolved to fear?', *Cognition & Emotion*, 2013, 27(7), pp. 1168–80

15 M. Mori, K. F. MacDorman and N. Kageki, 'The uncanny valley [from the field]', *Robotics & Automation Magazine, IEEE*, 2012, 19(2), pp. 98–100

16 M. E. Bouton and R. C. Bolles, 'Contextual control of the extinction of conditioned fear', *Learning and Motivation*, 1979, 10(4), pp. 445–66

17 W. J. Magee et al., 'Agoraphobia, simple phobia, and social phobia in the National Comorbidity Survey', *Archives of General Psychiatry*, 1996, 53(2), pp. 159–68

18 L. H. A. Scheller, 'This Is What A Panic Attack Physically Feels Like', http://www.huffingtonpost.com/2014/10/21/panic-attack-feeling_n_5977998.html (accessed September 2015)

19 J. Knowles et al., 'Results of a genome-wide genetic screen for panic disorder', *American Journal of Medical Genetics*, 1998, 81(2), pp. 139–47

20 E. Witvrouw et al., 'Catastrophic thinking about pain as a predictor of length of hospital stay after total knee arthroplasty: a prospective study', *Knee Surgery, Sports Traumatology, Arthroscopy*, 2009, 17(10), pp. 1189–94

21 R. Lieb et al., 'Parental psychopathology, parenting styles, and the risk of social phobia in offspring: a prospective-longitudinal community study', *Archives of General Psychiatry*, 2000, 57(9), pp. 859–66

22 J. Richer, 'Avoidance behavior, attachment and motivational conflict', *Early Child Development and Care*, 1993, 96(1), pp. 7–18

23 http://www.nhs.uk/conditions/social-anxiety/Pages/Social-anxiety.aspx (accessed September 2015)

24 G. F. Koob, 'Drugs of abuse: anatomy, pharmacology and function of reward pathways', *Trends in Pharmacological Sciences*, 1992, 13, pp. 177–84

25 L. Reyes-Castro et al., 'Pre-and/or postnatal protein restriction in rats impairs learning and motivation in male offspring', *International Journal of Developmental Neuroscience*, 2011, 29(2), pp. 177–82

26 W. Sluckin, D. Hargreaves and A. Colman, 'Novelty and human aesthetic preferences', *Exploration in Animals and Humans*, 1983, pp. 245–69

27 B. C. Wittmann et al., 'Mesolimbic interaction of emotional valence and reward improves memory formation', *Neuropsychologia*, 2008, 46(4), pp. 1000–1008

28 A. Tinwell, M. Grimshaw and A. Williams, 'Uncanny behaviour in survival horror games', *Journal of Gaming & Virtual Worlds*, 2010, 2(1), pp. 3–25

29 See Chapter 2, n. 29

30 R. S. Neary and M. Zuckerman, 'Sensation seeking, trait and state anxiety, and the electrodermal orienting response', *Psychophysiology*, 1976, 13(3), pp. 205–11

31 L. M. Bouter et al., 'Sensation seeking and injury risk in downhill skiing', *Personality and Individual Differences*, 1988, 9(3), pp. 667–73

32 M. Zuckerman, 'Genetics of sensation seeking', in J. Benjamin, R. Ebstein and R. H. Belmake (eds), *Molecular Genetics and the Human Personality*, Washington, DC, American Psychiatric Association, pp. 193–210.

33 S. B. Martin et al., 'Human experience seeking correlates with hippocampus volume: Convergent evidence from manual tracing and

voxel-based morphometry', *Neuropsychologia*, 2007, 45(12), pp. 2874–81
34 R. F. Baumeister et al., 'Bad is stronger than good', *Review of General Psychology*, 2001, 5(4), p. 323
35 S. S. Dickerson, T. L. Gruenewald and M. E. Kemeny, 'When the social self is threatened: Shame, physiology, and health', *Journal of Personality*, 2004, 72(6), pp. 1191–216
36 E. D. Weitzman et al., 'Twenty-four hour pattern of the episodic secretion of cortisol in normal subjects', *Journal of Clinical Endocrinology & Metabolism*, 1971, 33(1), pp. 14–22
37 See n. 12, above
38 R. S. Nickerson, 'Confirmation bias: A ubiquitous phenomenon in many guises', *Review of General Psychology*, 1998, 2(2), p. 175

4 Think you're clever, do you?

1 R. E. Nisbett et al., 'Intelligence: new findings and theoretical developments', *American Psychologist*, 2012, 67(2), pp. 130–59
2 H.-M. Süß et al., 'Working-memory capacity explains reasoning ability – and a little bit more', *Intelligence*, 2002, 30(3), pp. 261–88
3 L. L. Thurstone, *Primary Mental Abilities*, University of Chicago Press, 1938
4 H. Gardner, *Frames of Mind: The Theory of Multiple Intelligences*, Basic Books, 2011
5 A. Pant, 'The Astonishingly Funny Story of Mr McArthur Wheeler', 2014, http://awesci.com/the-astonishingly-funny-story-of-mr-mcarthur-wheeler/ (accessed September 2015)
6 T. DeAngelis, 'Why we overestimate our competence', *American Psychological Association*, 2003, 34(2)
7 H. J. Rosen et al., 'Neuroanatomical correlates of cognitive self-appraisal in neurodegenerative disease', *Neuroimage*, 2010, 49(4), pp. 3358–64
8 G. E. Larson et al., 'Evaluation of a "mental effort" hypothesis for correlations between cortical metabolism and intelligence', *Intelligence*, 1995, 21(3), pp. 267–78
9 G. Schlaug et al., 'Increased corpus callosum size in musicians', *Neuropsychologia*, 1995, 33(8), pp. 1047–55
10 E. A. Maguire et al., 'Navigation-related structural change in the hippocampi of taxi drivers', *Proceedings of the National Academy of Sciences*, 2000, 97(8), pp. 4398–403
11 D. Bennabi et al., 'Transcranial direct current stimulation for memory enhancement: From clinical research to animal models', *Frontiers in Systems Neuroscience*, 2014, issue 8

12 Y. Taki et al., 'Correlation among body height, intelligence, and brain gray matter volume in healthy children', *Neuroimage*, 2012, 59(2), pp. 1023–7
13 T. Bouchard, 'IQ similarity in twins reared apart: Findings and responses to critics', *Intelligence, Heredity, and Environment*, 1997, pp. 126–60
14 H. Jerison, *Evolution of the Brain and Intelligence*, Elsevier, 2012
15 L. M. Kaino, 'Traditional knowledge in curricula designs: Embracing indigenous mathematics in classroom instruction', *Studies of Tribes and Tribals*, 2013, 11(1), pp. 83–8
16 R. Rosenthal and L. Jacobson, 'Pygmalion in the classroom', *Urban Review*, 1968, 3(1), pp. 16–20

5 Did you see this chapter coming?

1 R. C. Gerkin and J. B. Castro, 'The number of olfactory stimuli that humans can discriminate is still unknown', edited by A. Borst, *eLife*, 2015, 4 e08127; http://www.ncbi.nlm.nih.gov/pmc/articles/PMC4491703/ (accessed September 2015)
2 L. Buck and R. Axel, 'Odorant receptors and the organization of the olfactory system', *Cell*, 1991, 65, pp. 175–87
3 R. T. Hodgson, 'An analysis of the concordance among 13 US wine competitions', *Journal of Wine Economics*, 2009, 4(01), pp. 1–9
4 See Chapter 1, n. 8
5 M. Auvray and C. Spence, 'The multisensory perception of flavor', *Consciousness and Cognition*, 2008, 17(3), pp. 1016–31
6 http://www.planet-science.com/categories/experiments/biology/2011/05/how-sensitive-are-you.aspx (accessed September 2015)
7 http://www.nationalbraille.org/NBAResources/FAQs/ (accessed September 2015)
8 H. Frenzel et al., 'A genetic basis for mechanosensory traits in humans', *PLOS Biology*, 2012, 10(5)
9 D. H. Hubel and T. N. Wiesel, 'Brain Mechanisms of Vision', *Scientific American*, 1979, 241(3), pp. 150–62
10 E. C. Cherry, 'Some experiments on the recognition of speech, with one and with two ears', *Journal of the Acoustical Society of America*, 1953, 25(5), pp. 975–9
11 D. Kahneman, *Attention and Effort*, Citeseer, 1973
12 B. C. Hamilton, L. S. Arnold and B. C. Tefft, 'Distracted driving and perceptions of hands-free technologies: Findings from the 2013 Traffic Safety Culture Index', 2013
13 N. Mesgarani et al., 'Phonetic feature encoding in human superior temporal gyrus', *Science*, 2014, 343(6174), pp. 1006–10

14 See Chapter 3, n. 14
15 D. J. Simons and D. T. Levin, 'Failure to detect changes to people during a real-world interaction', *Psychonomic Bulletin & Review*, 1998, 5(4), pp. 644–9
16 R. S. F. McCann, D. C. Foyle and J. C. Johnston, 'Attentional Limitations with Heads-Up Displays', *Proceedings of the Seventh International Symposium on Aviation Psychology*, 1993, pp. 70–5

6 Personality: a testing concept

1 E. J. Phares and W. F. Chaplin, *Introduction to Personality* (4th edn), Prentice Hall, 1997
2 L. A. Froman, 'Personality and political socialization', *Journal of Politics*, 1961, 23(02), pp. 341–52
3 H. Eysenck and A. Levey, 'Conditioning, introversion-extraversion and the strength of the nervous system', in V. D. Nebylitsyn and J. A. Gray (eds), *Biological Bases of Individual Behavior*, Academic Press, 1972, pp. 206–20
4 Y. Taki et al., 'A longitudinal study of the relationship between personality traits and the annual rate of volume changes in regional gray matter in healthy adults', *Human Brain Mapping*, 2013, 34(12), pp. 3347–53
5 K. L. Jang, W. J. Livesley and P. A. Vemon, 'Heritability of the big five personality dimensions and their facets: A twin study', *Journal of Personality*, 1996, 64(3), pp. 577–92
6 M. Friedman and R. H. Rosenman, *Type A Behavior and Your Heart*, Knopf, 1974
7 G. V. Caprara and D. Cervone, *Personality: Determinants, Dynamics, and Potentials*, Cambridge University Press, 2000
8 J. B. Murray, 'Review of research on the Myers-Briggs type indicator', *Perceptual and Motor Skills*, 1990, 70(3c), pp. 1187–1202
9 A. N. Sell, 'The recalibrational theory and violent anger', *Aggression and Violent Behavior*, 2011, 16(5), pp. 381–9
10 C. S. Carver and E. Harmon-Jones, 'Anger is an approach-related affect: evidence and implications', *Psychological Bulletin*, 2009, 135(2), pp. 183–204
11 M. Kazén et al., 'Inverse relation between cortisol and anger and their relation to performance and explicit memory', *Biological Psychology*, 2012, 91(1), pp. 28–35
12 H. J. Rutherford and A. K. Lindell, 'Thriving and surviving: Approach and avoidance motivation and lateralization', *Emotion Review*, 2011, 3(3), pp. 333–43
13 D. Antos et al., 'The influence of emotion expression on perceptions

of trustworthiness in negotiation', *Proceedings of the Twenty-fifth AAAI Conference on Artificial Intelligence*, 2011

14 S. Freud, *Beyond the Pleasure Principle*, Penguin, 2003

15 S. McLeod, 'Maslow's hierarchy of needs', *Simply Psychology*, 2007 (updated 2014), http://www.simplypsychology.org/maslow.html (accessed September 2015)

16 R. M. Ryan and E. L. Deci, 'Self-determination theory and the facilitation of intrinsic motivation, social development, and well-being', *American Psychologist*, 2000, 55(1), p. 68

17 M. R. Lepper, D. Greene and R. E. Nisbett, 'Undermining children's intrinsic interest with extrinsic reward: A test of the "overjustification" hypothesis', *Journal of Personality and Social Psychology*, 1973, 28(1), p. 129

18 E. T. Higgins, 'Self-discrepancy: A theory relating self and affect', *Psychological Review*, 1987, 94(3), p. 319

19 J. Reeve, S. G. Cole and B. C. Olson, 'The Zeigarnik effect and intrinsic motivation: Are they the same?', *Motivation and Emotion*, 1986, 10(3), pp. 233–45

20 S. Shuster, 'Sex, aggression, and humour: Responses to unicycling', *British Medical Journal*, 2007, 335(7633), pp. 1320–22

21 N. D. Bell, 'Responses to failed humor', *Journal of Pragmatics*, 2009, 41(9), pp. 1825–36

22 A. Shurcliff, 'Judged humor, arousal, and the relief theory', *Journal of Personality and Social Psychology*, 1968, 8(4p1), p. 360

23 D. Hayworth, 'The social origin and function of laughter', *Psychological Review*, 1928, 35(5), p. 367

24 R. R. Provine and K. Emmorey, 'Laughter among deaf signers', *Journal of Deaf Studies and Deaf Education*, 2006, 11(4), pp. 403–9

25 R. R. Provine, 'Contagious laughter: Laughter is a sufficient stimulus for laughs and smiles', *Bulletin of the Psychonomic Society*, 1992, 30(1), pp. 1–4

26 C. McGettigan et al., 'Individual differences in laughter perception reveal roles for mentalizing and sensorimotor systems in the evaluation of emotional authenticity', *Cerebral Cortex*, 2015, 25(1) pp. 246–57

7 Group hug!

1 A. Conley, 'Torture in US jails and prisons: An analysis of solitary confinement under international law', *Vienna Journal on International Constitutional Law*, 2013, 7, p. 415

2 B. N. Pasley et al., 'Reconstructing speech from human auditory cortex', *PLoS-Biology*, 2012, 10(1), p. 175

3 J. A. Lucy, *Language Diversity and Thought: A Reformulation of the Linguistic Relativity Hypothesis*, Cambridge University Press, 1992
4 I. R. Davies, 'A study of colour grouping in three languages: A test of the linguistic relativity hypothesis', *British Journal of Psychology*, 1998, 89(3), pp. 433–52
5 O. Sacks, *The Man Who Mistook His Wife for a Hat, and Other Clinical Tales*, Simon and Schuster, 1998
6 P. J. Whalen et al., 'Neuroscience and facial expressions of emotion: The role of amygdala–prefrontal interactions', *Emotion Review*, 2013, 5(1), pp. 78–83
7 N. Guéguen, 'Foot-in-the-door technique and computer-mediated communication', *Computers in Human Behavior*, 2002, 18(1), pp. 11–15
8 A. C.-y. Chan and T. K.-f. Au, 'Getting children to do more academic work: foot-in-the-door versus door-in-the-face', *Teaching and Teacher Education*, 2011, 27(6), pp. 982–5
9 C. Ebster and B. Neumayr, 'Applying the door-in-the-face compliance technique to retailing', *International Review of Retail, Distribution and Consumer Research*, 2008, 18(1), pp. 121–8
10 J. M. Burger and T. Cornelius, 'Raising the price of agreement: Public commitment and the lowball compliance procedure', *Journal of Applied Social Psychology*, 2003, 33(5), pp. 923–34
11 R. B. Cialdini et al., 'Low-ball procedure for producing compliance: commitment then cost', *Journal of Personality and Social Psychology*, 1978, 36(5), p. 463
12 T. F. Farrow et al., 'Neural correlates of self-deception and impression-management', *Neuropsychologia*, 2015, 67, pp. 159–74
13 S. Bowles and H. Gintis, *A Cooperative Species: Human Reciprocity and Its Evolution*, Princeton University Press, 2011
14 C. J. Charvet and B. L. Finlay, 'Embracing covariation in brain evolution: large brains, extended development, and flexible primate social systems', *Progress in Brain Research*, 2012, 195, p. 71
15 F. Marlowe, 'Paternal investment and the human mating system', *Behavioural Processes*, 2000, 51(1), pp. 45–61
16 L. Betzig, 'Medieval monogamy', *Journal of Family History*, 1995, 20(2), pp. 181–216
17 J. E. Coxworth et al., 'Grandmothering life histories and human pair bonding', *Proceedings of the National Academy of Sciences*, 2015. 112(38), pp. 11806–11
18 D. Lieberman, D. M. Fessler and A. Smith, 'The relationship between familial resemblance and sexual attraction: An update on Westermarck, Freud, and the incest taboo', *Personality and Social Psychology Bulletin*, 2011, 37(9), pp. 1229–32

19 A. Aron et al., 'Reward, motivation, and emotion systems associated with early-stage intense romantic love', *Journal of Neurophysiology*, 2005, 94(1), pp. 327–37
20 A. Campbell, 'Oxytocin and human social behavior', *Personality and Social Psychology Review*, 2010
21 W. S. Hays, 'Human pheromones: have they been demonstrated?', *Behavioral Ecology and Sociobiology*, 2003, 54(2), pp. 89–97
22 L. Campbell et al., 'Perceptions of conflict and support in romantic relationships: The role of attachment anxiety', *Journal of Personality and Social Psychology*, 2005, 88(3), p. 510
23 E. Kross et al., 'Social rejection shares somatosensory representations with physical pain', *Proceedings of the National Academy of Sciences*, 2011, 108(15), pp. 6270–75
24 H. E. Fisher et al., 'Reward, addiction, and emotion regulation systems associated with rejection in love', *Journal of Neurophysiology*, 2010, 104(1), pp. 51–60
25 J. M. Smyth, 'Written emotional expression: Effect sizes, outcome types, and moderating variables', *Journal of Consulting and Clinical Psychology*, 1998, 66(1), p. 174
26 H. Thomson, 'How to fix a broken heart', *New Scientist*, 2014, 221(2956), pp. 26–7
27 R. I. Dunbar, 'The social brain hypothesis and its implications for social evolution', *Annals of Human Biology*, 2009, 36(5), pp. 562–72
28 T. Dávid-Barrett and R. Dunbar, 'Processing power limits social group size: computational evidence for the cognitive costs of sociality', *Proceedings of the Royal Society of London B: Biological Sciences*, 2013, 280(1765), 10.1098/rspb.2013.1151
29 S. E. Asch, 'Studies of independence and conformity: I. A minority of one against a unanimous majority', *Psychological Monographs: General and Applied*, 1956, 70(9), pp. 1–70
30 L. Turella et al., 'Mirror neurons in humans: consisting or confounding evidence?', *Brain and Language*, 2009, 108(1), pp. 10–21
31 B. Latané and J. M. Darley, 'Bystander "apathy"', *American Scientist*, 1969, pp. 244–68
32 I. L. Janis, *Groupthink: Psychological Studies of Policy Decisions and Fiascoes*, Houghton Mifflin, 1982
33 S. D. Reicher, R. Spears and T. Postmes, 'A social identity model of deindividuation phenomena', *European Review of Social Psychology*, 1995, 6(1), pp. 161–98
34 S. Milgram, 'Behavioral study of obedience', *Journal of Abnormal and Social Psychology*, 1963, 67(4), p. 371
35 S. Morrison, J. Decety and P. Molenberghs, 'The neuroscience of group membership', *Neuropsychologia*, 2012, 50(8), pp. 2114–20

36 R. B. Mars et al., 'On the relationship between the "default mode network" and the "social brain"', *Frontiers in Human Neuroscience*, 2012, vol. 6, article 189
37 G. Northoff and F. Bermpohl, 'Cortical midline structures and the self', *Trends in Cognitive Sciences*, 2004, 8(3), pp. 102–7
38 P. G. Zimbardo and A. B. Cross, *Stanford Prison Experiment*, Stanford University, 1971
39 G. Silani et al., 'Right supramarginal gyrus is crucial to overcome emotional egocentricity bias in social judgments', *Journal of Neuroscience*, 2013, 33(39), pp. 15466–76
40 L. A. Strömwall, H. Alfredsson and S. Landström, 'Rape victim and perpetrator blame and the just world hypothesis: The influence of victim gender and age', *Journal of Sexual Aggression*, 2013, 19(2), pp. 207–17

8 When the brain breaks down . . .

1 V. S. Ramachandran and E. M. Hubbard, 'Synaesthesia – a window into perception, thought and language', *Journal of Consciousness Studies*, 2001, 8(12), pp. 3–34
2 See Chapter 3, n. 1
3 R. Hirschfeld, 'History and evolution of the monoamine hypothesis of depression', *Journal of Clinical Psychiatry*, 2000
4 J. Adrien, 'Neurobiological bases for the relation between sleep and depression', *Sleep Medicine Reviews*, 2002, 6(5), pp. 341–51
5 D. P. Auer et al., 'Reduced glutamate in the anterior cingulate cortex in depression: An in vivo proton magnetic resonance spectroscopy study', *Biological Psychiatry*, 2000, 47(4), pp. 305–13
6 A. Lok et al., 'Longitudinal hypothalamic–pituitary–adrenal axis trait and state effects in recurrent depression', *Psychoneuroendocrinology*, 2012, 37(7), pp. 892–902
7 H. Eyre and B. T. Baune, 'Neuroplastic changes in depression: a role for the immune system', *Psychoneuroendocrinology*, 2012, 37(9), pp. 1397–416
8 W. Katon et al., 'Association of depression with increased risk of dementia in patients with type 2 diabetes: The Diabetes and Aging Study', *Archives of General Psychiatry*, 2012, 69(4), pp. 410–17
9 A. M. Epp et al., 'A systematic meta-analysis of the Stroop task in depression', *Clinical Psychology Review*, 2012, 32(4), pp. 316–28
10 P. F. Sullivan, M. C. Neale and K. S. Kendler, 'Genetic epidemiology of major depression: review and meta-analysis', *American Journal of Psychiatry*, 2007, 157(10), pp. 1552–62
11 T. H. Holmes and R. H. Rahe, 'The social readjustment rating scale',

Journal of Psychosomatic Research, 1967, 11(2), pp. 213–18

12 D. H. Barrett et al., 'Cognitive functioning and posttraumatic stress disorder', *American Journal of Psychiatry*, 1996, 153(11), pp. 1492–4

13 P. L. Broadhurst, 'Emotionality and the Yerkes–Dodson law', *Journal of Experimental Psychology*, 1957, 54(5), pp. 345–52

14 R. S. Ulrich et al., 'Stress recovery during exposure to natural and urban environments' *Journal of Environmental Psychology*, 1991, 11(3), pp. 201–30

15 K. Dedovic et al., 'The brain and the stress axis: The neural correlates of cortisol regulation in response to stress', *Neuroimage*, 2009, 47(3), pp. 864–71

16 S. M. Monroe and K. L. Harkness, 'Life stress, the "kindling" hypothesis, and the recurrence of depression: Considerations from a life stress perspective', *Psychological Review*, 2005, 112(2), p. 417

17 F. E. Thoumi, 'The numbers game: Let's all guess the size of the illegal drug industry', *Journal of Drug Issues*, 2005, 35(1), pp. 185–200

18 S. B. Caine et al., 'Cocaine self-administration in dopamine D3 receptor knockout mice', *Experimental and Clinical Psychopharmacology*, 2012, 20(5), p. 352

19 J. W. Dalley et al., 'Deficits in impulse control associated with tonically-elevated serotonergic function in rat prefrontal cortex', *Neuropsychopharmacology*, 2002, 26, pp. 716–28

20 T. E. Robinson and K. C. Berridge, 'The neural basis of drug craving: An incentive-sensitization theory of addiction', *Brain Research Reviews*, 1993, 18(3), pp. 247–91

21 R. Brown, 'Arousal and sensation-seeking components in the general explanation of gambling and gambling addictions', *Substance Use & Misuse*, 1986, 21(9–10), pp. 1001–16

22 B. J. Everitt et al., 'Associative processes in addiction and reward the role of amygdala-ventral striatal subsystems', *Annals of the New York Academy of Sciences*, 1999, 877(1), pp. 412–38

23 G. M. Robinson et al., 'Patients in methadone maintenance treatment who inject methadone syrup: A preliminary study', *Drug and Alcohol Review*, 2000, 19(4), pp. 447–50

24 L. Clark and T. W. Robbins, 'Decision-making deficits in drug addiction', Trends in Cognitive Sciences, 2002, 6(9), pp. 361–3

25 M. J. Kreek et al., 'Genetic influences on impulsivity, risk taking, stress responsivity and vulnerability to drug abuse and addiction', *Nature Neuroscience*, 2005, 8(11), pp. 1450–57

26 S. S. Shergill et al., 'Functional anatomy of auditory verbal imagery in schizophrenic patients with auditory hallucinations', *American Journal of Psychiatry*, 2000, 157(10), pp. 1691–3

27 P. Allen et al., 'The hallucinating brain: a review of structural and

functional neuroimaging studies of hallucinations' *Neuroscience & Biobehavioral Reviews*, 2008, 32(1), pp. 175–91
28 S.-J. Blakemore et al., 'The perception of self-produced sensory stimuli in patients with auditory hallucinations and passivity experiences: evidence for a breakdown in self-monitoring', *Psychological Medicine*, 2000, 30(05), pp. 1131–9
29 See n. 27, above
30 R. L. Buckner and D. C. Carroll, 'Self-projection and the brain', *Trends in Cognitive Sciences*, 2007, 11(2), pp. 49–57
31 A. W. Young, K. M. Leafhead and T. K. Szulecka, 'The Capgras and Cotard delusions', *Psychopathology*, 1994, 27(3–5), pp. 226–31
32 M. Coltheart, R. Langdon, and R. McKay, 'Delusional belief', *Annual Review of Psychology*, 2011, 62, pp. 271–98
33 P. Corlett et al., 'Toward a neurobiology of delusions', *Progress in Neurobiology*, 2010, 92(3), pp. 345–69
34 J. T. Coyle, 'The glutamatergic dysfunction hypothesis for schizophrenia', *Harvard Review of Psychiatry*, 1996, 3(5), pp. 241–53

Index